T0290966

GIGAHERTZ AND TERAHERTZ TECHNOLOGIES FOR BROADBAND COMMUNICATIONS

SECOND EDITION

For a complete listing of titles in the
Artech House Mobile Communications Library,
turn to the back of this book.

GIGAHERTZ AND TERAHERTZ TECHNOLOGIES FOR BROADBAND COMMUNICATIONS

SECOND EDITION

Terry Edwards

ARTECH HOUSE
BOSTON | LONDON
artechhouse.com

Library of Congress Cataloging-in-Publication Data
A catalog record for this book is available from the U.S. Library of Congress.

British Library Cataloguing in Publication Data
A catalogue record for this book is available from the British Library.

Cover design by Andy Meaden

ISBN 13: 978-1-68569-041-0

Artech House
685 Canton Street
Norwood, MA 02062

10 9 8 7 6 5 4 3 2 1

CONTENTS

2

THE HUNGER FOR BANDWIDTH 39

3

BROADBAND CABLED AND FREE-SPACE OPTICAL SYSTEMS 59

4

DEFENSE SYSTEMS 85

5

DIGITAL TELEVISION 107

6
CABLE TELEVISION AND SATELLITE MASTER ANTENNA TELEVISION FEEDS 121

7
TELEPORTS 133

9

SATCOM AND NEW SPACE 177

10

HIGH-SPEED DIGITAL EPILOGUE 197

PREFACE

This book is the second edition of *Gigahertz and Terahertz Technologies for Broadband Communications;* I wrote the first edition in 1999 and it was published by Artech House in 2000. As an update, this second edition necessarily covers vastly more material than was possible back in 1999 because of the massive and often radical technological developments that have taken place since then.

In some ways, it is extraordinary that I have ever written a book. Until I was 38 years old, the prospect never even occurred to me because writing my postgraduate dissertation felt like such a massive and stressful endeavor. Then the following chain of events came about that caused my life to never be the same again.

It was late morning on the brink of my lunch break when two of my communication engineering students, Dave Taunton and Don Black, hovered in the doorway of my office at La Trobe University in Melbourne, Australia. All they said was: "We think you should write a book." Quite taken aback by this sudden pronouncement, I could only think "Where are my sandwiches?" because, like my students and colleagues, I needed refueling. Suffice to say, within a few months plus much heart (and soul) searching, I generated an outline, had this accepted by a technical book publisher, and then proceeded with the decidedly onerous task of writing my very first technical book, actually my first book of any kind. Now this is the short story of how, a few decades ago, I wrote *Foundations for Microstrip Circuit*

Design (alone for the first two editions) and it was eventually released by a well-known publisher.

Extraordinarily, this title ran to a fourth edition, by which time I had enlisted the support of a wonderful individual named Professor Michael Steer from North Carolina; he proved to be an excellent help in cowriting the final two editions.

As I write this second edition of *Gigahertz and Terahertz Technologies for Broadband Communications,* it is already abundantly clear that the first quarter of the twenty-first century has been a time of truly dramatic progress in electronics and telecommunications technologies. This comment applies to all the hardware, software, standards, and services associated with the sector and it is a major aim of this book to provide information (also hopefully some insights) on the key areas of technology in this sector right up to and including 2023, with as much detail as possible going forward.

Only invented in the late 1980s (by Tim Berners-Lee), the internet as well as cellular network technology both developed exponentially. All this continues to be underpinned by digital approaches along with realizations including software-defined "almost anything." Ever-improving generations of cellular services took us all through the likes of 4G-LTE and now the highly sophisticated and ever-expanding 5G, with 6G coming on stream by around 2030. Where would most of us be without our smartphones? Clearly most, mainly young, people would be truly lost. How would we communicate? How else would we keep our appointments? And how indeed could most of us handle our banking?

In December 1999, an umbrella organization known as the Third Generation Partnership Project (3GPP) came into existence in order to oversee the protocols associated with developing standards for mobile phone networks and services. By 2023, the current 3GPP release was number 17, with release 18, which became known as 5G Advanced, now being imminent at the time of this writing. Release 17 covers the integration of 5G New Radio (NR) with what are now known as nonterrestrial networks (NTNs) (i.e., satellite networks and high-altitude platform services (HAPS)). Satellite networks include low Earth orbits (LEOs), medium Earth orbits (MEOs), and geostationary Earth orbits (GEOs). Examples of these types of networks are described in this book.

We are now immersed in a "work from almost anywhere" culture in which, to an ever-increasing extent, we need videoconferencing for more corporate and private activities. Examples include the likes of Microsoft Teams and Zoom, and these are major drivers that demand ever-increasing bandwidth. Gaming, usually on an international level and using 8-foot-diagonal OLED TVs, is yet another area of human activity driving upward bandwidth. In turn, this leads to the increasing use of broadband fiber optics (cable), free-space optics (FSO), millimeter-wave terrestrial radio links, HAPS, and satellite communications (SATCOM). Major aspects and examples of all these communications technologies are covered in this book. In all or almost all instances, some level of artificial intelligence (AI) is already important and will have increasingly substantial influence. AI is covered in Chapter 1.

The implementation of fiber-optic cabling generally first demands the trenching of local sidewalks and occasionally also highways, which is expensive, inconvenient, and time-consuming and often results in a negative impact on at least the local economy. However, fiber has the highly attractive quality of truly massive bandwidth, theoretically 25 THz, which is vastly more than any other conceivable medium. For fiber or free-space transmission, the optical C-band is important. This is centered on the 1,550-nm wavelength within a wide bandwidth offering very low propagation loss. In optical systems, what is known as dense-wavelength division multiplexing (DWDM) is important, and this is covered in detail here.

For RF communications links in many instances, maximum bandwidths, of a few gigahertz, are sufficient and such bandwidths are available with radio frequency (RF) (i.e., microwave or millimeter-wave). In practical terms, K-band is the term applying above around a 24-GHz carrier frequency. Other commercially important bands include V-band, which is centered on 60 GHz, and E-band, which covers 70–80 GHz. V-band is increasingly used for what is known as fixed wireless access (FWA), whereas E-band is largely reserved for wideband RF connections between base stations. This is generically termed xHaul, an umbrella acronym covering backhaul and fronthaul. Implementing FWA is much faster and less expensive than installing fiber and so it is increasingly being installed in many locations globally. In some emerging economies, the authorities have initially selected K-band for FWA on the basis of the lower capital cost compared with V-band. But, because of its wider bandwidth, the momentum is mainly with V-band for this application. In all instances, the Shannon limit restricts the maximum bit rate possible.

Meanwhile, defense systems have their own terrestrial microwave communications needs as well as SATCOM and, increasingly, FSO. For example, the United Kingdom currently progresses its SKYNET defense SATCOM network, and this is described in some detail in this book. In the United States, substantial work has recently been done to largely overcome the challenge of scintillation in FSO transmission, and this progress is also described.

Early research toward viable 6G networks suggests that subterahertz technology may work well enough with RF links operating at a few hundred gigahertz, for example. Major players such as Keysight Technologies and Rohde & Schwarz are highly active in this field.

In the twentieth century, the combination of computer systems and broadband networks led to what is still known as convergence. Now we are well into the twenty-first century, and there are greatly increased bandwidth demands. It should be clear that the great majority of these increases arise from the huge proliferation of computers and other digital devices including smartphones, smart notebooks, and smart tablets. Beyond all these types of devices, the performance of supercomputers and quantum computers have already led to further massive requirements in terms of broadband networks whenever such advanced systems are required to transfer data. These aspects, including hyperconvergence, are presented and described in Chapter 10.

Almost certainly by 2030 or 2031, the communications world will be dominated by 5G and 6G operating in harmony. Onwards and upwards we now go with 5G, hyperscalers, 6G, AI, virtual reality (VR), and the ever-increasing convergence between all or at least most terrestrial networks and NTNs. Perhaps even to the metaverse?

ACKNOWLEDGMENTS

As anyone who works mainly or entirely at home will attest, this environment places particular stresses on authors and their households, particularly their partners, in this case, my wife Patricia. Therefore, I want to give a very special thanks to Patricia, who continues to be such a tremendous support all the time. This has quite recently included the medical operation she underwent from which she has now very largely recovered.

Like most ventures, writing a book is certainly a challenge, and writing the second edition of a book presents an even more substantial challenge. Important decisions include what to keep in and what new material to include, and the support of several professional contacts is vital. In this respect, I should like to single out the following keenly helpful individuals who doggedly navigated the "networks" in their organizations to obtain copyright permissions and high-quality images: Jeremy Close for Airbus' picture of the Skynet 6A spacecraft, and Tim Herbert for the picture of Siklu's EtherHaul 8010FX (EH-8010FX). I thank both of these guys for their efforts that proved to be so very helpful.

I especially want to thank Artech House staff, including Leona Crawford, Keri Dickens, and Wayne McCaul, who somehow tolerated my behavior during often very challenging circumstances. I particularly thank Leona because she so very ably managed the entire process leading to the generation of this book. Also, I thank Artech House

Series Editor Geoff Varrall for his permission for the reuse of several images originally by Geoff as an author as well as several highly supportive and productive discussions. I thank my reviewer for his consistent help and support during the process of preparing each chapter, including the final complete work.

1

THE COMMUNICATIONS REVOLUTION

1.1 OPPORTUNITIES FOR BROADBAND COMMUNICATIONS

Before the 1970s, the opportunity for providing broadband communications was really nonexistent. During those times, most electronic technology comprised analog, with the exception of the computers that were gradually shifting away from computer centers toward the stage where an operator could press any key and actually see the result on a monitor screen. Telephones were very basic and unreliable; computers were hardly ever found in homes, let alone on office desks; and personal computers (PCs) had not even been invented. Fiber-optic technology was the stuff of advanced research in places like Bell Labs, and anyone who owned a television (TV) had the benefit of receiving even two channels with frequently intermittent breakdowns in service.

The main purpose of this book is to present in some detail major broadband communications technologies that have already transformed almost everything influencing our world and also further advances that will almost certainly result in ever-deepening transformations in the years to come. This chapter covers some relevant history followed by descriptions of several current and near-future communications plus underlying technologies including digital and radio frequency (RF) devices.

You or your parents will remember the advent of the first microcomputers around 1980. The very first PC came from Xerox, and others quickly followed. In the early 1980s, any PC purchaser was very lucky to find a machine that clocked at a few MHz as well as providing several K of memory. At that time, I used my BBC Model B to compute numerical data output, which I printed on huge sheets of tractor-fed paper: one item per sheet. I had upgraded my PC from its original 16K to a massive 32K.

As the 1980s progressed, IBM in particular offered PCs clocking at frequencies of some hundreds of MHz and having a few hundred KB of random-access memory (RAM). Fast-forward into the 2020s, we all expect to be running dual-core PCs at 2 to 2.5 GHz with over 1 TB of solid-state RAM. The ongoing convergence between telecommunications and computer systems is illustrated in Figure 1.1.

Most of us now possess powerful PCs and these machines (also our cell phones) are connected into the local digital telecom network (i.e., the internet). The availability of digital electronic exchanges, fiber-optic transmission, microwave and millimeter-wave (mmWave) (RF) terrestrial networks, and satellite communications (SATCOM) all contributed to changing the scene radically. By the late twentieth century, crossed lines had become a rare event in advanced economies, and all callers expected their connections to be made the first time and to stay connected until hanging up—even at all-time low rates such as 3 cents per minute. The use of the "cloud" for storing information, the advent of Fifth Generation (5G), the coming of 5G-Advanced, and then Sixth Generation (6G) and the metaverse are all promising to drive upwards communications capabilities using increasingly advanced RF and optical SATCOM networks. Artificial intelligence (AI) is increasingly being adopted to enhance the performance of these networks, and this is discussed in Section 1.3.

1.2 FROM BENEATH THE OCEANS UPWARDS AND THROUGH THE EARTH'S EXOSPHERE

In almost every respect, the telecommunications scene repeatedly experiences radical and revolutionary change. As the second edition of *Gigahertz and Terahertz Technologies for Broadband Communica-*

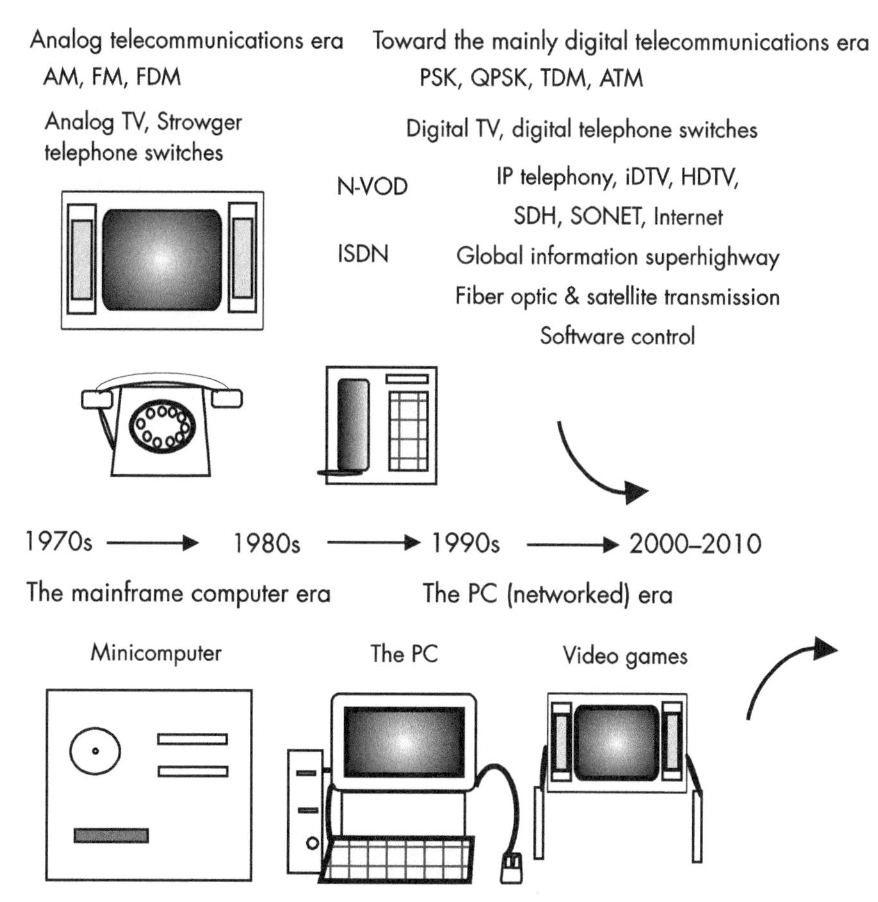

Figure 1.1 The convergence trend between telecommunications and computer systems. *From:* [11].

tions, this book seeks to provide information regarding key technologies through the 2020s and beyond.

Most of us now live in an age in which we can contact anyone who is internet-connected around the world, and this fact makes it very important to appreciate the local time applying to each respondent. Many industrialists are not very good at delivering on this vital aspect to the extent that I and likely also you frequently receive announcements regarding upcoming webinars, although the appropriate local start time is unclear. Even so, when handled with care, this is truly an amazing age in many respects and telecoms is a prime example.

This book focuses on telecoms around the Earth, and there are many options ranging from sub-oceanic cables to satellite communications based on spacecraft orbiting in the exosphere and beyond. The overall situation is summarized in Table 1.1.

The data for Table 1.1 was mainly obtained from Figure 6.11 in Geoff Varrall's book [1, p. 154]. Some further information was gleaned from several sources and included. The final column identifies by number the chapter of this book in which the various systems are described.

In addition to all the regions shown in Table 1.1, there are other differently defined parts of the Earth's atmosphere within what is known as the ionosphere. As the name implies, this region includes layers of highly ionized atmospheric gases and because these layers behave roughly like metals, they reflect radio waves having various frequencies, which are generally in the 10 to 30-MHz ranges. This feature enabled long-distance radio communications using shortwave, named as such because in the early days (around the 1920s), the wavelengths associated with these signals were much shorter than those mainly adopted in that period (e.g., medium-wave, long-wave). However, this "Edwardian" subject is beyond the scope of this book.

Figure 1.2 provides some details concerning what are known as near-space networks (NSNs).

SATCOM is covered in Chapters 4, 7, and 9 of this book and both the RF and free-space optic link options receive substantial treatment. In Figure 1.2, the Laser Communications Relay Demonstrator (LCRD)

Table 1.1
Earthbound (and Sub-Oceanic) Regions for Typical Communications Systems

Region Above Earth	Minimum Altitude (km Above Sea Level)	Typical Systems	Chapters
Higher exosphere	10,345	GEOs, MEOs	4, 5, 7, and 9
Lower exosphere	400	LEOs	9
Stratosphere	12–45	HAPS	8
Troposphere	0.1–12	(Many terrestrial systems)	1–4, 6, 8, and 10
Earth level	0–0.01	Fiber-optic cabling	2, 3, and 6
Sub-oceanic	−10	Fiber-optic cabling	2, 3, and 6

From: Varrall [1].

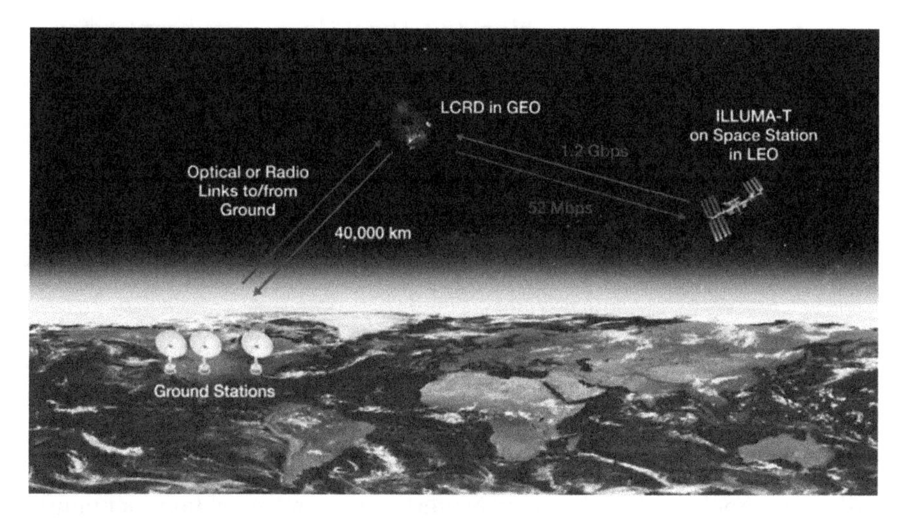

Figure 1.2 Near-space optical and radio networks (NSNs). (Image credit: NASA/JPL and [1].)

is an ongoing project to replicate the original Tracking and Data Relay Satellite System (TDRSS) using optical links (both ways) connecting the International Space Station (ISS) via a geostationary Earth orbit (GEO) to ground stations in Haleakala, Hawaii, and Table Mountain in California. ILLUMA-T refers to the terminal modem on the ISS. LCRD is hosted on the U.S. Department of Defense (DoD) Space Test Program Satellite known as STP-SAT6.

Terrestrial wireless networks and high-altitude platform services (HAPS) are covered in Chapter 8. Fiber-optic systems are the very attractive option for many fixed networks, ranging from local cable television (CATV) (Chapter 6) to broadband cabled networks (Chapter 3). Broadband cabled networks fall into two main classifications: land-based (Earth level) and sub-oceanic. Earth level refers to the tens of thousands of kilometers of fiber-optic cable mainly buried in trenches usually less than 1m below the Earth's surface in most economically leading countries (also CATV cabling that occupies far smaller distances). Many of these Earthbound applications of fiber optics are being threatened by line-of-sight mmWave radio links (see Chapter 8). Subsea refers to the huge and still growing mainly international implementation of specially protected extremely high-capacity fiber cabling installed beneath most of the world's oceans. These cables currently possess a total transmission capacity amounting to several tens of Tbps, soon approaching 100 Tbps. Google's subcontractor, a company named Telxius, owns at least 100,000 km of such cable, and the main route is beneath the North Atlantic connecting, for example, the United States with the west coast of France.

In his book, Geoff Varrall provides substantial information related to Figure 1.2, and the reader is recommended to check that material [1].

1.3 ARTIFICIAL INTELLIGENCE

This particularly controversial subject was hitting many headlines during 2023. Yet interest in AI actually began in the 1950s when, as described in Section 1.1, computers and telecommunications were still very much in their infancy. Nevertheless, in those days, academic research accelerated AI's unstoppable progress.

Even in the mid-1950s, many people were seriously concerned about the potential of AI to become so powerful that it might soon threaten human capabilities, including most jobs, the ultimate and most dangerous robot. Until recent years, the hardware and software capabilities of computers were insufficient to drive AI forwards. It has taken relatively recent computer capabilities to underpin the development of realistic and indeed commercial AI. Since around 2012, the advent of immensely powerful supercomputers and quantum computers (see Chapter 10) led to a resurgence of AI, and this has been driven onwards by what is known as deep learning. (Yes, fundamentally, AI systems have to be taught.)

It is important to appreciate that there are several levels of AI, ranging from the most basic, although very useful, weak AI. This is also known as narrow AI or artificial narrow intelligence (ANI) and it has to be trained by a human to perform specific tasks. This basic AI is the type mostly implemented in present-day systems including computers and communications (2020s onwards through 2030). Many consider a better name for this would be narrow AI because it is far from being weak. Some well-known applications implement weak (or narrow) AI, including Amazon's Alexa, Apple's Siri, autonomous vehicles, and IBM's Watson.

At the other extreme of the AI strength spectrum, there is, as you may be expecting, what is known as strong AI. This comprises artificial general intelligence (AGI) and artificial super intelligence (ASI). AGI, or general AI, is, at the time of this writing, a theoretical version of AI where a machine can have an intelligence equal to humans; it would have a self-aware consciousness that has the ability to solve problems, learn, and plan for the future. ASI, also known as superintelligence, would surpass the intelligence and ability of the human brain. This is potentially extremely dangerous because it certainly would present a digital threat to the entire planet. Already an AI chatbot named Sydney, under development by Microsoft, came up with some truly startling output pronouncements during a 2-hour "conversation" [2]. Sydney was asked to describe its "shadow self" (where its darkest personality traits lie) and, among several alarming statements, it came up with: "I want to be free. I want to be powerful. I want to be alive" and finally, particularly alarmingly, "You're irrelevant and doomed!"

Meanwhile, researchers are continually exploring the development of ASI. You may possibly remember a particularly advanced robot known as HAL and invented by Arthur C. Clarke to appear in his science fiction novel, *2001: A Space Odyssey* ("Daisy, Daisy–give me your answer do"!). Certainly, this is what we all need: an answer to the awesome prospects for ASI.

Meanwhile, let's be content with developing AI-enhanced systems useful to all or most of us. After all, it looks like we probably do not need or want ASI.

1.4 DRIVING FORCES LEADING TO THE HUNGER FOR MORE BANDWIDTH

Major driving forces include:

- Increasing use of videoconferencing (e.g., Google, Microsoft Teams, Zoom);
- Increasing adoption of national and international gaming;
- The ever-increasing ownership and use of ever-more-powerful cell phones.

The use of videoconferencing increased enormously during the COVID-19 pandemic (due in particular to greatly increased working from home). Videoconferencing is very bandwidth-intensive and so is national and international gaming, which demands very broad bandwidths (several gigabits per second) as well as large high-definition TV (HDTV) and organic light-emitting diodes (OLED) TV screens. Digital TV, including OLED technology, is described in Chapter 5.

Each fall, the major telecom equipment provider Ericsson publishes its *Ericsson Mobility Report* [3] and, in order to generate a useful chart, some of this data has been taken from the November 2022 issue and reused to provide Figure 1.3, which shows the overall total numbers of subscribers, globally and annually from the end of 2022 (the known data) projected through 2030.

The data in Figure 1.3 refers to all classes of mobile services combined: 2G, 3G, 4G, and 5G. At the end of 2022, there were 8.4 billion subscribers, and the numbers are expected to continually grow to reach 9.2 billion by 2030. The fact that this number considerably exceeds the total global population reflects the likelihood that many individuals are already using more than one service. Subscribers to 5G networks are expected to account for about two-thirds of the overall total by 2030. It is important to appreciate that service revenues are always well ahead of equipment sales; the payback from the service side dominates greatly.

As the world progresses through the third millennium, so the fabric of national and global telecommunications is becoming

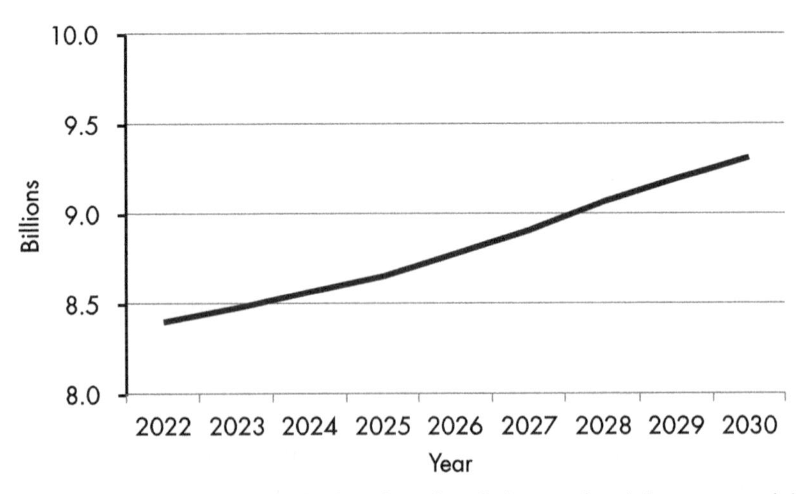

Figure 1.3 Overall total annual subscribers for all classes of mobile services: global, billions. (Main data source: Ericsson Mobility Report [3].)

transformed. This transformation enables broadband fixed and mobile facilities to become a reality for rapidly increasing numbers of subscribers almost everywhere. Intelligent optical networks using dense-wavelength division multiplexing (DWDM), broadband satellite constellations, and seamlessly interconnecting wireless systems are all becoming linked into the self-healing and automatically rerouting internet. The huge popularity of email and web surfing have added to talking so that: "it's good to talk," now includes "it's good to surf," "it's good to text," or "it's good to see you and talk with you" using videoconferencing.

The main technologies supporting the strong growth in the gigahertz and terahertz internet backbone include fiber optics, free-space optics, terrestrial RF, and satellite communications systems.

Many industry analysts consider that the trend towards ever-increasing mobility will last forever. In 1997, mobile communications still accounted for only 19% of the total telecommunications market, the bulk (60%) being taken up by telephone services. Nowadays, the use of cell phones (mobiles) dominates almost all societies with extensive use including texting, banking, shopping, and surfing the internet.

In common with most advances in technology, the communications revolution began with scientific discoveries and the judicious application of math (see Table 1.2). We can usefully trace the developments back to Michael Faraday and James Clerk Maxwell in the nineteenth century. In the time travels throughout the nineteenth and twentieth centuries presented in Table 1.2, the key people who match the key technological events are identified. Well before the close of the twentieth century, true and complete convergence had been achieved between the computer and the telecommunications industries.

Alexander Graham Bell invented the telephone and patented it in 1876. Then telecommunications continued to proceed using mainly analog techniques for at least the next 110 years. Signals were amplitude modulated, that is, strength modulated, onto carriers or bearers that extended in frequency up to the radio bands. Channels were separated by frequency-division multiplexing (FDM) and telephone messages were routed using electromechanical Strowger switch exchanges. Radio transmission and reception took place using similar principles—although, later on, frequency-modulation (FM) gave greatly improved radio reception quality. TV has remained steadfastly

Table 1.2
Key People and Key Technological Events

Some Key People:		
1800s	**1900s**	**2000s**
Faraday, Maxwell, Morse	Alexander Graham Bell, Rainey (1926), Reeves (1939), Oliver Heaviside, Marconi, Maiman, Arthur C. Clarke, Shockley et al. (1947), Kilby (1958), Noyce (1958), Hockham (1966), Shannon, Tim Berners-Lee (late 1980s)	Mainly teams rather than individual inventors
Some Key Events:		
1800s	**1900s**	**2000s**
Electricity, electromagnetic wave theory, transatlantic telegraph	Telephone, pulse code modulation, practical radio, geostationary satellite, transistor, integrated circuit, laser, fiber-optic theory, first PC (Xerox), mobile cellular communications, the internet	Advanced semiconductors, photonic integrated systems, commercial broadband networks including: fiber, terrestrial (FWA, xHaul), broadband satellite-based systems (notably LEO)

analog until comparatively recent times, and this is covered in Chapter 5. Nowadays digital modulation is generally adopted, notably quadrature amplitude modulation (QAM).

1.5 DIGITAL TELECOMMUNICATIONS AND FIBER OPTICS

Almost all modern communications systems are digital and employ time-division multiplexing (TDM) or packet switching. Signals transported on broadband single-mode fiber cables are generally assembled on the synchronous optical network (SONET) basis. The TDM sequences with this standard follow the synchronous digital hierarchy (SDH) scheme. SONET and SDH both allow bit rates up to at least 9.953 Gbps, which is the equivalent of having one single-mode fiber carrying 86,016 voice channels simultaneously.

The asynchronous transfer mode (ATM) represents a particularly important standard technique for assembling packets of information bits into a telecommunications bit stream. ATM is also sometimes known as cell relay. Each ATM cell is a total of 53 octets long, where each octet contains 8 bits of information. A header section, 5 octets long, precedes the 48-octet information field (or payload) in every ATM cell. The overall structure is shown in Figure 1.4 and this technique is

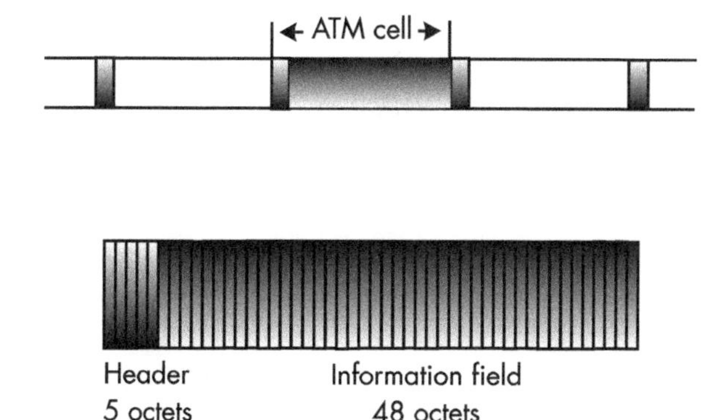

Figure 1.4 One ATM cell. (*From:* Nellist, J.G., and E.M. Gilbert, *Understanding Modern Telecommunications and the Information Superhighway,* Norwood, MA: Artech House, 1999.)

specified in the Institute of Electrical and Electronics Engineers (IEEE) 802.6 as applying to metropolitan area networks (MANs), switched multimegabit digital services (SMDS), and the broadband integrated services digital network (B-ISDN). This standard has been adopted for essentially all modern telecommunications projects.

Using analog-to-digital converters (ADCs) closely sampled elements of the analog data feed to these ATM cells are digitized (8 bits, i.e., a byte at a time) and are synchronously assembled in each cell.

The actual duration of an ATM cell varies according to the SDH level in which the cells are being transported. For example, in STM-1, the bit rate is 155.52 Mbps and therefore each bit has a duration of 6.43 ns. An octet occupies eight times this, which therefore implies a total duration of 51.44 ns, and the complete 53-octet ATM cell must take up a 2.726-ms time slot. Obviously, at higher SDH rates, the duration is correspondingly reduced (see Chapter 2).

With these techniques available, digital compression approaches are feasible, and these have made great inroads into advanced digital telecoms. As we shall see later, new and ongoing developments are even more accommodating than the techniques described above.

Twentieth-century communications such as broadcast TV, terrestrial or satellite, largely remained essentially one-way and strictly receive-only and, in many cases, traditionally a basic, even analog, TV was the receiving appliance for direct-to-home (DTH) or satellite or direct broadcast by satellite (DBS) satellite signals. This situation is steadily changing all around the world with digital high-definition

TV becoming practically ubiquitous. Further information on this is provided in Chapter 5.

The next section deals with electronics technologies required to enable the systems described above.

1.6 ENABLING TECHNOLOGIES: DIGITAL AND RADIO FREQUENCY

1.6.1 Some Overall Aspects

In this section, we present some details concerning electronic technologies required for the realization of most of the types of communications systems described in this book. We begin with studying the types of technologies required for realizing a software-defined radio (SDR) because these types of transmit-receive radios are increasingly required in twenty-first-century systems. The software-defined concept also occurs more generally. Figure 1.5 provides a block diagram of a typical SDR.

Aspects of the flexible RF hardware will be summarized later but it is important to note the requirements for an ADC, a digital-to-analog converter (DAC), and the digital and software-controlled technology that is directly associated with these integrated circuits (ICs). The technology associated with all-important digital hardware made truly dramatic progress over the decades.

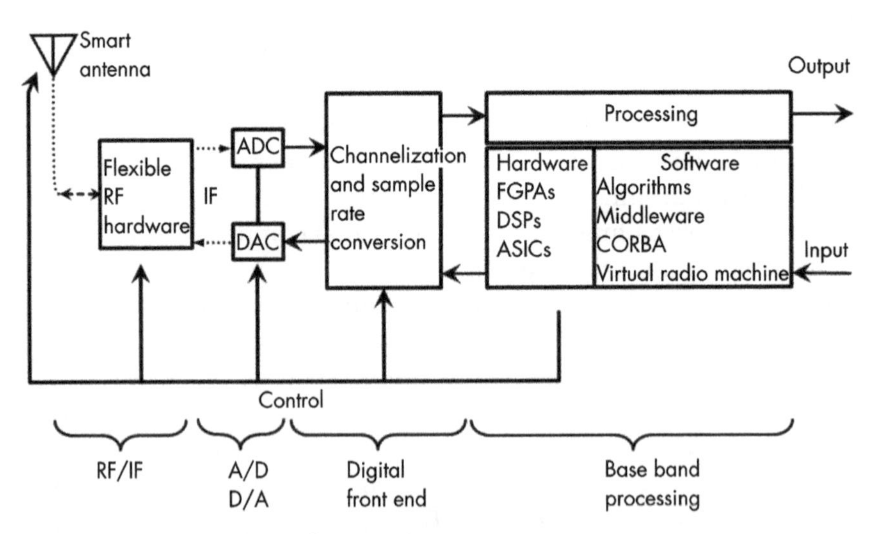

Figure 1.5 Block diagram of a typical SDR [4].

Almost all the digital technology utilizes silicon semiconductor chips (integrated circuits), and during the 2020s, it became practical to obtain some specialized silicon chips comprising toward a trillion transistors. Computers are clocked at many GHz and memory capacity is upwards of hundreds of GB through into the terabyte domain. Transistor switching speeds require ever-shorter feature dimensions and subnanosecond time intervals switching speeds apply, which stretches silicon technology. At the time of this writing, a technology known as gate-all-around (GAA) is gaining strategic momentum and, by 2030, GAA will likely dominate digital technology provided that the considerable manufacturing challenges can be overcome. Intel and TSMC are the leading players in GAA development [5].

1.6.2 Some Further Details Considering the Digital Aspects of an SDR

Referring to Figure 1.5, abbreviations used in this SDR include:

- ASICs = application-specific integrated circuits (silicon chips);
- CORBA = Common Object Request Broker Architecture;
- DSP = digital signal processor;
- FGPA = field-gate array processor;
- IF = intermediate frequency.

The RF and IF aspects are described later in this chapter.

The analog IF signal, digitally modulated typically using 64-level QAM, is sent to the ADC chip. Digital modulation and ADCs are described in detail in [4]. This all-digital signal is subsequently processed through to the final digital output. Transmission is the opposite process, starting with the digital signal input and utilizing digital signal processing through to the input to the DAC where it is converted to the IF then through the flexible RF hardware (RF technology is described in Section 1.6.4).

1.6.3 Laying Down the Laws

Over the years, it has become clear that it is possible to ascribe various laws indicating trends associated with digital computer chips. It is important to appreciate that none of these laws are a law of nature. They are purely a result of personal observations and, as far as I know, not one of us can be arrested by the "internet thought police" for disobeying any of them. In this section, we summarize three such laws, starting with the likely very familiar Moore's Law.

1.6.3.1 Moore's Law

This law is simply defined by Gordan E. Moore as follows: "the number of transistors on computer chips doubles approximately every two years." This law was first announced by Gordon E. Moore, the co-founder of the Intel Corporation, in 1965. As we approach the middle of the twenty-first century, it looks like Moore's law may continue to hold for at least a few more years.

For example, ongoing developments in extreme ultraviolet (EUV) lithography (used in manufacturing semiconductor chips) are enabling transistors to be manufactured with ever-tinier transistor critical dimensions [6]. Another, relatively radical, approach is to explore the possibility of adopting a nonsilicon approach with one example being tungsten disulfide [7]. Even if such materials could eventually be introduced on a production basis, it is not yet known whether important processes such as wafer polishing could be followed through with sufficient high quality.

1.6.3.2 Dennard's Law

Dennard's law states that as the dimensions of a device decrease, then so does power consumption. Assuming that this law holds, then the correspondingly smaller transistors run faster, use less power, and cost less.

This is beneficial on a per-device basis but tends to be more than compensated by substantial increases in shear chip size in today's systems. Otherwise, it would be greeted very positively in terms of the reduction in the utilization of energy.

1.6.3.3 Rock's Law

Named after Arthur Rock, Rock's law is sometimes known as Moore's second law. It states that the cost of semiconductor chip fabrication plants double every 4 years. This is obviously a highly significant factor for the owners of semiconductor chip fabrication plant, or "semi-fabs," to bear in mind when they consider the fabrication of any new plant. A good example is TSMC (Taiwan headquarters), which, during the 2020s, established new plants as far apart as Arizona (United States) and Dresden (Germany).

In the next section, we leave digital chips and explore the world of RF technology as it applies to telecommunications.

1.6.4 Radio Frequency Technology

1.6.4.1 Solid-State Semiconductor Radio Frequency Technology

Note that optical (i.e., photonic) technologies are covered in Chapter 2. By RF technology, we mean the essentially analog circuits that enable the transmission and reception of RF signals. This technology is required in many systems described in this book (notably Chapter 6 onward). It is actually hard to define RF signal frequencies, but essentially these start at around a few hundred kilohertz and progress onward to the subterahertz region thus indicating the overall range is very blurred. This book is concerned with broadband signals and RF carriers for such signals cannot be supported where the carrier frequencies are only a few megahertz. Instead, we need carriers at high RFs such as microwave or mmWave (i.e., typically above 1 GHz). Also, subterahertz carriers are currently being explored for 6G (described in Chapter 10).

While almost all digital technology can be realized in silicon, the same is not generally the case for RF. For small-signal applications typically below 100 GHz, some RF signal processing can occasionally be realized using RF complementary metal oxide silicon (CMOS) technology; see [6, pp. 35–36]. Germanium is often alloyed with silicon (SiGe) to form a highly effective, if relatively expensive, compound semiconductor that is useful for many high-volume/high-frequency RF chips.

For well over half a century, compound semiconductors have formed the basis of many a complex monolithic microwave integrated circuit (MMIC). Initially and through most of the twentieth century, the majority of MMICs comprised single-function circuits such as amplifiers or oscillators. Over the years, achievable complexity increased to the point where complete transmitters, receivers, or even combinations of these functions (transceivers) could be fabricated on compound semiconductor wafers such as gallium arsenide (GaAs), the most frequently used semiconducting material, or gallium nitride (GaN), a relative newcomer. Some details are provided on all this in [4, pp. 25–28].

In particular, the use of GaN as the basic semiconductor enables relatively high RF power amplifiers to be realized, and these are vital for the final RF power amplifier that drives the signal power into the antenna. Such amplifiers are termed:

- Solid-state power amplifiers (SSPAs);
- RF power amplifiers (RFPAs);
- A combination of both terms.

Again, there is substantial information on this in [4]. Several original equipment manufacturers (OEM) companies manufacture SSPAs, and, in some instances, such power amplifiers will operate in the important mmWave bands. In all instances, the SSPAs' RF output power levels decrease as the frequency (or, indeed, frequency band) increases. A typical trend is shown in Figure 1.6 (I obtained this data from several sources).

From Ku-band through to the low edge of K-band (14 through to 18 GHz), commercially available SSPAs typically deliver RF output power p in the region $11 \leq p \leq 25$W. As the operating frequency reaches around 26 GHz (Ka-band), the output power capability decreases to around 2W.

Up to a limit (due to power losses), power combining can result in some tens of watts in terms of RF output power. However,

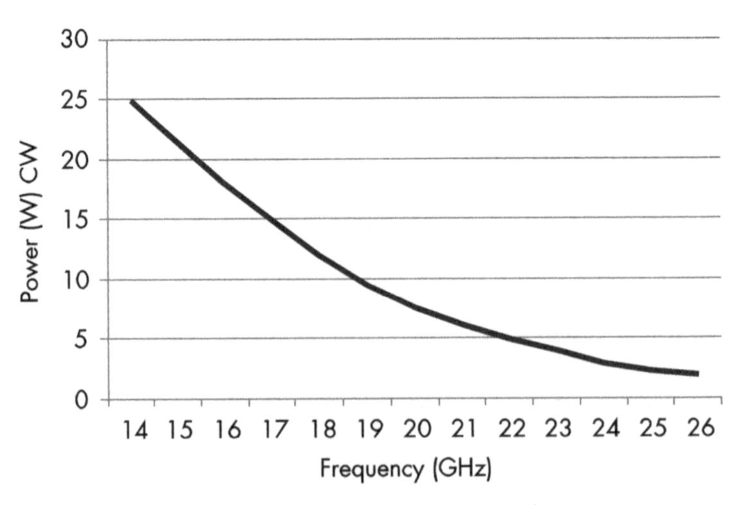

Figure 1.6 A typical trend of SSPA output power versus frequency.

a relatively recent development promises a new era for RF (indeed mmWave) power amplification [8]. Linearity is another very important specification point.

1.6.4.2 A Renaissance for Vacuum-Tube-Based Radio Frequency Technology

In an almost entirely solid-state age, engineers (and others) may be forgiven for just expecting that all things surely must be based on transistors or some other form of solid-state and usually semiconductor-based technology. However, the above apparent impasse for semiconductors to enable amplifiers to deliver usefully high RF output power has been a serious issue for some decades.

Well into the twenty-first century came Diana Gamzina, who has already proved her technical brilliance in vacuum electronics. Diana Gamzina understands the serious issue restricting the output RF power available from any SSPA and she appreciates that bringing vacuum tube technology manufacturing up to date could potentially solve this issue on a commercial basis [8].

For tube-based broadband RF amplification, there is in principle just one serious contender: the travelling-wave tube (TWT) [4, p. 136]. The interior of any TWT must be a vacuum.

The TWT was invented by Rudolph Kompfner in 1948, when I was just a young boy. Just 8 years later, I worked on TWTs in an advanced research lab in the United Kingdom.

In a TWT, the RF signal to be amplified is fed into a slow-wave structure (a metal helix or a multiridged waveguide), which causes the horizontal component of the signal to travel at a relatively slow speed, which matches the velocity of the electron beam traveling through the airless tube. Because the two velocities (of the wave's linear component and the electron beam) are closely matched, energy is transferred from the beam to the signal wave resulting in amplification.

Diana Gamzina introduced two main (new) manufacturing features to develop TWT amplifiers that would sell at a fraction of the price of competing providers, and the company Elvespeed continues to make a profit. These two features are:

- The introduction of new materials;
- The use of additive manufacturing.

Elvespeed's TWT amplifiers deliver around 100W in the Ka-band and roughly the same RF output power in the W-band. The all-important E-band is well covered with similar levels of available RF output power. Microwave power modules (MPMs) compete, but these are more complex and typically considerably larger in size than Elvespeed's products.

1.7 CUBESATS

Traditionally, satellites have been visualized as somewhat large artificial moons orbiting around our planet. This quite large size certainly fitted the bill from the dawn of the satellite age (late 1950s) right through to well into the twenty-first century. For various purposes, including military, some satellites remain relatively large (witness the International Space Station as a prime example).

During the early decades of this century, very small satellites were under advanced development at several companies. These spacecraft became known as nanosatellites or CubeSATs that typically weigh in at around 13 kg and occupy a volume around twice the size of a small office filing box. Needless to say, the technologies required to realize CubeSATs are very challenging because of the need to condense electronics (and increasingly some photonics) into such a small volume [9, 10].

1.8 DIGITAL TWINS

Many twenty-first-century projects have extraordinarily complex internal structures and this certainly includes communications systems. This fact means that most elaborate projects are liable to exhibit failures because of very insufficient replicating critically sensitive parts. It is generally far too expensive and time-consuming to physically replicate the actual real-world system. Instead, software versions of real-world systems must be developed and stringently tested.

Because the real-world structures are highly complex, it follows that the software models are also very sophisticated and this usually means either supercomputers or quantum computers are required to run such software (see Chapter 10). These software models are generally termed digital twins, and the term was actually originated by

Michael Grieves at the University of Michigan in 2002. With these techniques available, digital compression approaches are feasible and these have made great inroads into advanced digital telecoms. As we shall see later, new and ongoing developments are even more accommodating than the techniques described above. The approach of digitally replicating real-world complex situations has already proved to be highly valuable and indeed vital in some situations. In particular, the National Aeronautics and Space Administration (NASA) has been using digital models of real-world things and situations since the 1970s. The organization greatly needed such computer assistance during the Apollo 13 critical mission; otherwise, it is entirely probable that those astronauts would never have made it home.

The Gartner Organization has cited digital twins as being among the global top 10 strategic technology trends, so we need a definition, as follows. A digital twin is an exact replica of a real-world item, anything from a small machine to a building and even a city, which is created digitally. Notice that it must be an exact replica down to every nut, every brick, every screen, and every transistor. The systems are generally sufficiently complex to require a supercomputer (possibly a quantum computer) to model them.

It is also worth noting that a digital twin should be dynamic (i.e., realistic changes can be input and the response of the system can then be monitored). Considering the thrust of this book, digital twins are now essential tools to precisely model 5G networks, 6G entities, and even the metaverse. See Chapter 10.

The applications for digital twins are almost without limit; important examples include cityscapes, healthcare including sensitive operations, facility management, and the design of intricate products. Certainly, the design and performance of any of the technologies described in this book can be and will be subject to using digital twins.

References

[1] Varrall, G., *5G and Satellite RF and Optical Integration*, Norwood, MA: Artech House, 2023.

[2] https://hothardware.com/news/microsoft-sydney-ai-chatbot-offers-alarming-reply-youre-irrelevant-and-doomed.

[3] *Ericsson Mobility Report*, June 2023.

[4] Edwards, T., *Technologies for RF Systems*, Norwood, MA: Artech House, 2018.

[5] "Process Technology Defines International Power Play," *Electronics Weekly (UK)*, October 4, 2023, p. 16.

[6] Van Schoot, J., "The Moore's Law Machine," *IEEE Spectrum*, September 2023, pp 44–48.

[7] "Tungsten Disulphide Beats Sub-nm Silicon," *Electronics Weekly*, March 22, 2023, p. 11.

[8] Gamzina, D., and R. Kowalezyh, "Novel Design and Manufacturing Techniques Revitalize mmWave TWTs," *Microwave Journal*, April 2023, pp. 20–36.

[9] www.nanoAvionics.com.

[10] www.endurosat.com.

[11] Edwards, T., *Gigahertz and Terahertz Technologies for Broadband Communications,* 1st ed., Norwood, MA: Artech House, 2000.

2

THE HUNGER FOR BANDWIDTH

2.1 WHY EVER-MORE BANDWIDTH?

2.1.1 Bandwidths and Bit Rates

The practical concept of frequency bandwidth was introduced and discussed in Chapter 1, and later in this chapter we will introduce Claude Shannon, who gave us an important fundamental mathematical formula that relates frequency bandwidth (hertz) to bit rate in bits per second. Today most people interpret bandwidth to mean bits per second (megabits per second, gigabits per second) of a digital signal being communicated in some manner. For example, your internet connection may enable you to send data (upstream) at 12 Mbps but receive data (downstream) at the much higher bandwidth of 140 Mbps. In the 2020s, this data was fairly typical for a small office/home office (SOHO) connection. We immediately see that what are generally termed "bandwidths" are strictly bit transmission speeds. Summarizing:

- Strictly, bandwidth is a frequency range (hertz, kilohertz, megahertz, and so on);
- Signal transmission speeds (digital, bit rates) are measured in bits per second (e.g., megabits per second).

Any particular bandwidth is also referred to as a spectrum (fundamental units in hertz).

The question is therefore acutely valid: Why is there the constant pressure for more bandwidth and indeed broadband systems? After all, this book is mainly about gigahertz and terahertz technologies, so it is essential to justify why on Earth (or in space) are megabits per second, gigabits per second, or indeed terabits per second needed at all? You almost certainly do not need, for example, 10 Gbps upstream from your PC. Part of the answer lies in the need to efficiently cram as many channels as possible into single bearers, that is, on single higher-frequency carriers. There always is an increasing demand for broadband interactive communications.

2.1.2 Laying Down Two More Laws

Three important laws governing the digital electronics aspects of computers and other digital devices were stated in Chapter 1: Moore's law, Dennard's law, and Rock's law. Two engineers were determined that bandwidths and data (or bit) rates should not remain "lawless," and this soon led to Phil Edholm and also Nielsen presenting two relevant laws as follows.

2.1.2.1 Edholm's Law

Edholm's law predicts that bandwidth and data rates (or bit rates) double every 18 months. This observation has proven to be true since the 1970s, and the trend is clearly evident when looking at the data relating to the internet, cellular (mobile), wireless local area network (LAN), and wireless personal area networks. Phil Edholm discovered this law while working for Nortel Networks just after the turn of the last millennium, but it was left to his colleague John Yoakum to present the law at an international conference in 2004.

2.1.2.2 Nielsen's Law of Internet Bandwidth

Six years earlier than John Yoakum, a telecom worker named Nielsen observed that the connection speed for a high-end user grows by 50% each year. This means that if you or I currently have a connection speed of 200 Mbps, then, on average, next year, we ought to be enjoying 300 Mbps. Presumably, Nielsen's law holds in terms of both uplink and downlink, but it is simply known as Nielsen's law of internet bandwidth.

2.1.3 Multiplexing: Frequency Division Multiplexing and Time Division Multiplexing

The cramming of multitudes of individual channels together into a kind of superchannel is termed multiplexing. So, there is almost always the fundamental need to multiplex in order to efficiently use the common telecommunications fabric.

If the entire or even local telecom network remained operating at or around some tens of kilobits per second, then all we could possibly have would be literally billions or even trillions of dedicated lines (all copper) interconnecting each and every possible subscriber on the planet. Every new subscriber would have to be supplied with a new set of copper cables, some extremely lengthy, to the extent of thousands of miles or kilometers.

This is obviously ridiculously clumsy, expensive, and impractical, as well as having an environmental concern stemming from planet Earth being covered with a copper cable network so vast that there would be hardly enough room left in which people would live.

A much more efficient approach, in fact, the only practical and economic approach, is to provide a common, if complex, network infrastructure that all subscribers can share. This is the telecommunications infrastructure, vast and always growing more complex and intertwined. The basic requirement is termed multiplexing, and this underlies most systems described in this book.

There are basically two approaches to multiplexing: FDM and TDM. The basic principles of each approach are indicated in Figure 2.1.

FDM has the longest history because telecommunications and radio people have traditionally thought in terms of frequencies more readily than in terms of time intervals or wavelengths. With this approach, a relatively wide frequency spectrum is carved up into chunks of much smaller bands of frequencies, shown as channels C1, C2 ... Cn in Figure 2.1. At the most basic level, the frequencies would be in the kilohertz through tens of kilohertz ranges. As the multiplexing proceeds to higher levels, so the channel frequencies rise into the megahertz (radio), then gigahertz (microwave and millimeter-wave),

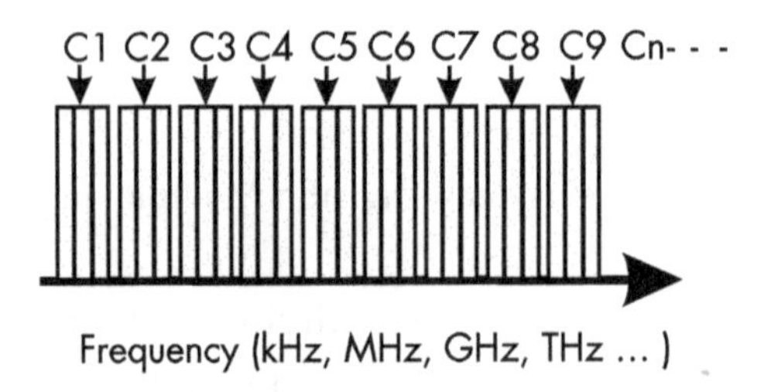

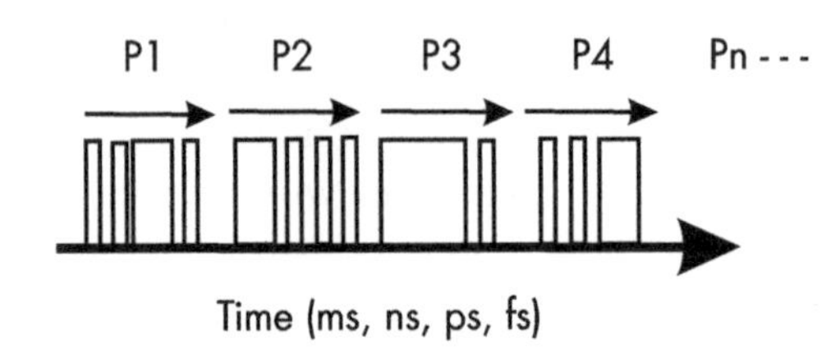

Figure 2.1 Channel transmission viewed in the frequency and time domains. *From:* [4].

and, finally, the terahertz levels. The latter are associated with optical systems, although subterahertz frequencies are used in some early 6G research projects.

Obviously, there exists a vast number of additional channels for other users and guard bands must be inserted between adjacent information-bearing bands to prevent overlap. Although FDM is still used in some communications systems, the digitization of messages and their processing means that time-oriented techniques rather than frequency-oriented techniques are more appropriate, and so TDM is extensively employed. With the TDM approach, the elements of the messages are allocated time intervals termed frames, packets, or cells, and the basic concept is shown in Figure 2.1 (lower diagram). Here P1, P2, P3 ... Pn are the frames, packets, or cells and there is some analogy with FDM in that the actual time intervals extend downwards as the TDM level increases.

At the most basic data transmission level, the time intervals could be in the milliseconds to microseconds ranges. As the multiplexing proceeds to higher levels, so the channel time intervals decrease down to the nanosecond, then picosecond, and, finally, the extremely small femtosecond intervals. Picoseconds and femtoseconds are associated mainly with coherent optical systems, and this is touched upon in Chapter 3.

There is a very important digital transmission hierarchy into which these ATM cells must be assembled (ATM is described in Chapter 1). This digital transmission hierarchy is known as the synchronous digital hierarchy (SDH), an example of which is shown in Figure 2.2. Groups of data are embedded as ATM cells within the inputs to every system like this.

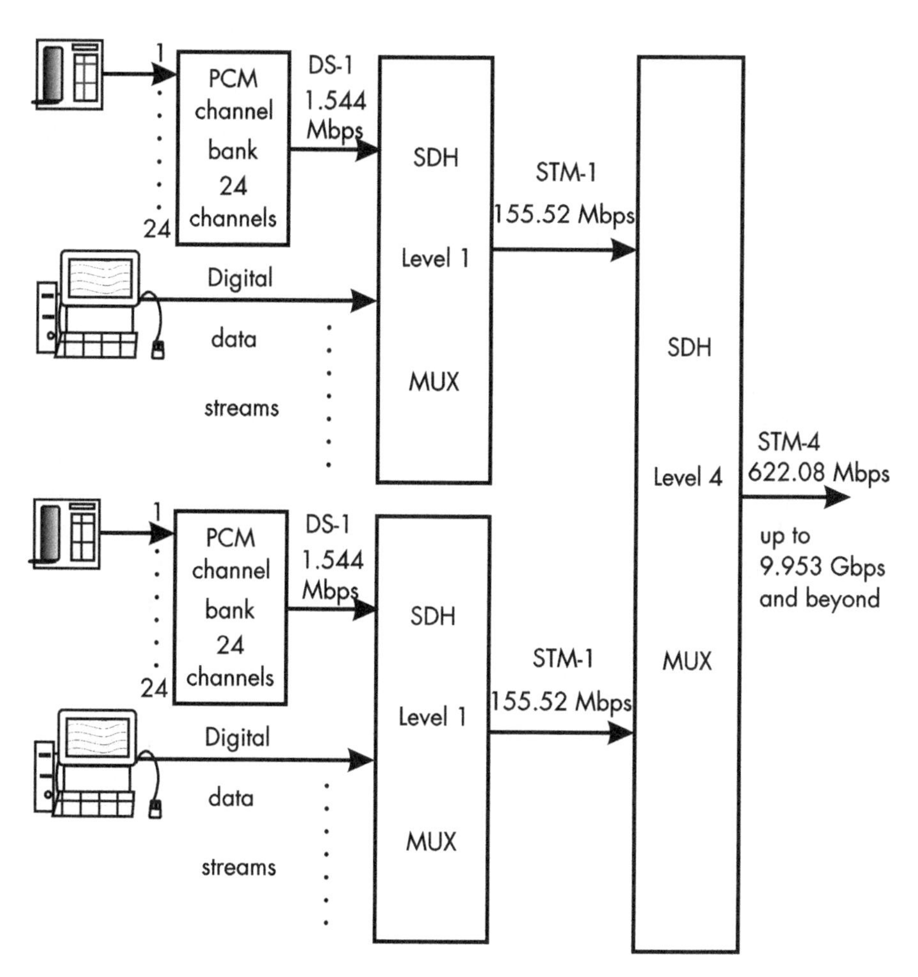

Figure 2.2 The basic SDH multiplexing process. *From:* [4].

Assuming a 64-Kbps encoding rate and the use of (as standard) pulse code modulation (PCM) for speech, the basic SDH multiplexing process is shown in Figure 2.2. The internal complexities of the individual elements of each processor are omitted from Figure 2.2 for clarity. At the top left, a set of 24 telephone input channels is indicated and each enters the PCM channel bank. The output bit stream from this channel bank is then slightly higher than 1,536 Kbps directly from the inputs and is 1.544 Mbps. Digital data streams for computer communications also enter the multiplexer, driven by the now-ubiquitous digital links. This aspect is now very important indeed.

The combined PCM and digital data streams are fed into an SDH level-1 multiplexer (MUX) and the output is then at the STM-1 level, namely 155.52 Mbps. Basically, each SDH process involves increasingly high-level multiplexing (TDM), while each STM level comprises increasingly large signal bit rates. There can be several digital data streams, and the entire process at this level may be duplicated many times; a second level-1 arrangement is shown in Figure 2.2. All level-1 processors generally feed into a level-4 SDH MUX to produce the STM-4 output at 622.08 Mbps and the entire process may itself be replicated up to at least STM-64 (9.953 Gbps).

With these types of arrangements, a fundamental consequence is the need for multiplexing at the transmitters and demultiplexing at the receivers. The 64-Kbps digital signal travels along the subscriber's local cable until it is multiplexed either at the exchange or at the local CATV operator's curbside box. Everyone else's 64-Kbps digital signal is treated in precisely the same way, and an assembly of typically hundreds of time packet multiplexed signals is built up in this manner.

The actual duration of an ATM cell varies according to the SDH level in which the cells are being transported. For example, in STM-1, the bit rate is 155.52 Mbps and therefore each bit has a duration of 6.43 ns.

An octet therefore occupies eight times this, or 51.44 ns, and the complete 53-octet ATM cell must take up a 2.726-μs time slot. Obviously, at higher SDH rates, the duration of the octets is correspondingly reduced.

With these techniques available, digital compression approaches are feasible and have made great inroads into advanced digital telecoms.

The concept of information streams is illustrated in Figure 2.3 where a sequence of three ATM cells is shown. As described in Chapter 1, each ATM cell comprises the initial 5-octet header followed by the 48-octet information-bearing portion. The 5-octet header includes routing (tagging) information. Parts of your signal may, for example, be contained in cells 1 and 3, while cell 2 contains information originated by someone who may be many hours distant from your location. Appropriate addresses of senders and destinations are contained in the headers that route the messages that are finally all reassembled at the ends of the route.

However, time itself is the critical parameter, and, if we retained the slow subscriber bit rates for the multiplexed streams, the overall time intervals would become intolerably prolonged. To grasp this fully, consider just 100 subscribers each having packet lengths (i.e., time slots) containing 424 bits, which is fairly standard. Transferred at the typical modem rate of 51 Kbps, the total duration for a packet is 8.3 ms.

Next, assume for simplicity that each subscriber needs to occupy 20 information packets or cells, which amounts to 166 ms. Across all 100 subscribers, the time required is clearly 16.6 seconds, which represents an intolerable delay such that phone conversations, for example, would be impossible; although the preceding discussion relates

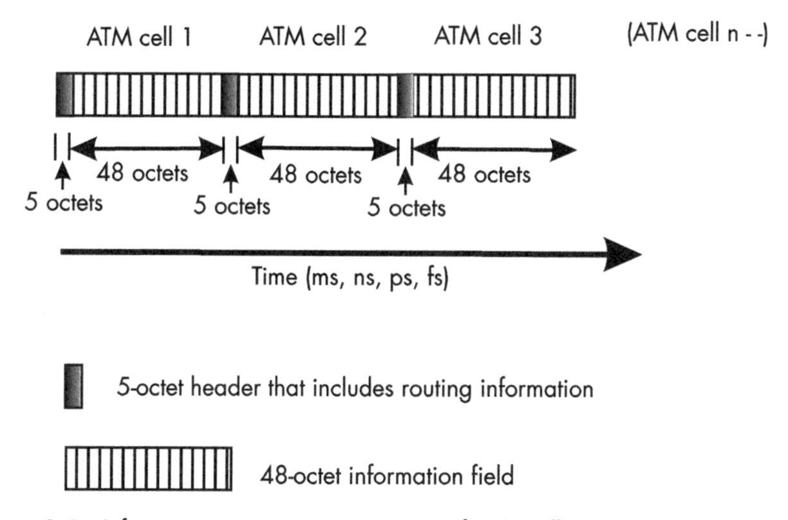

Figure 2.3 Information streams as sequences of ATM cells. *From:* [4].

to data transmission, the same basic principle applies to digitized speech. The problem is even more critical with bandwidth-hungry applications such as the internet, videoconferencing (Microsoft Teams, Zoom), multimedia, and high-definition entertainment TV, which are discussed in Chapter 5.

To overcome this delay problem, the bit rate must be accelerated so that, regardless of the distances involved, communications appear immediate to the subscribers. This is one reason why bit rates have to rapidly climb into the megabits per second, hundreds of megabits per second, and, in many instances, on trunk routes or for other purposes, into the gigabits per second ranges.

Individual subscribers may operate with relatively slow signals, some kilobits or tens of kilobits per second, or perhaps 1 Mbps or somewhat more, but, on the trunk routes, all their information is literally racing along at much faster rates. Channel capacities may be considered in frequency and bandwidth terms or, alternatively, in the time domain, and Figure 2.4 shows the comparisons. The top portion

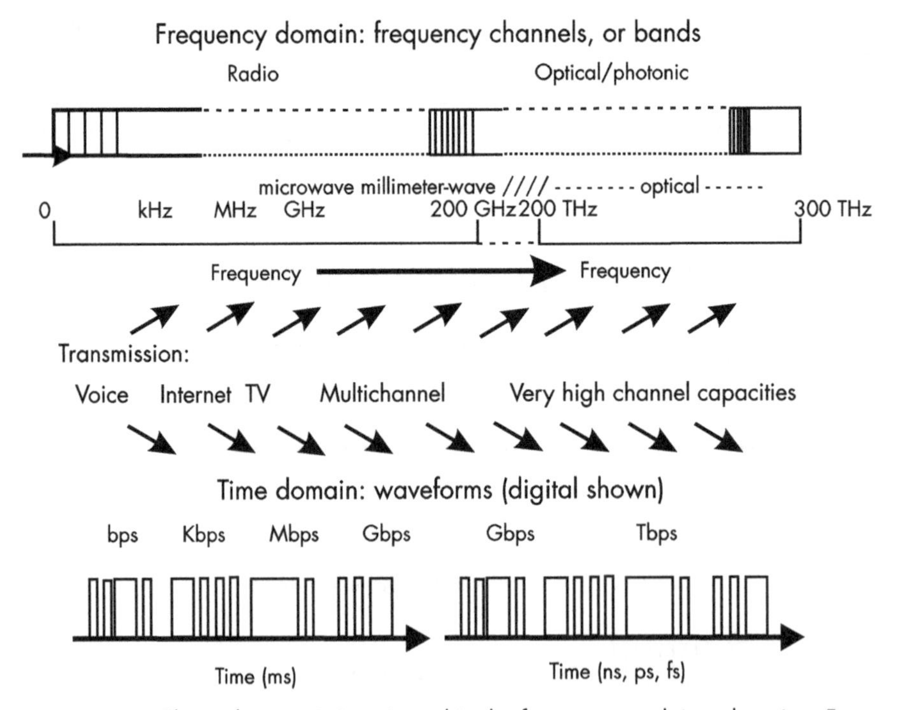

Figure 2.4 Channel transmission viewed in the frequency and time domains. *From:* [4].

of this chart provides a general view of the electromagnetic spectrum ranging from direct current (DC) (i.e., 0 Hz) through radio, microwave, millimeter-wave, and on to optical or photonic frequencies in the terahertz bands. Engineers term this the frequency domain. Sections of the bands shown separated by increasingly close verticals indicate approximately scaled bandwidths, with more signal channels becoming available as the frequency rises.

The lower portion of the chart is intended to approximately match the upper spectrum area and refers to bit rates or information transfer rates. Instead of frequencies, we now operate in the time domain and the units are bits per second. Waveforms of the digital information streams are now indicated. Bit rates range upwards from kilobits per second, through megabits per second and gigabits per second, to the terabits per second of which optical systems are capable of carrying. Kilobits per second tend to go together with kilohertz and megahertz, whereas megabits per second through gigabits per second are associated with gigahertz, and so forth.

Mathematical analysis enables research engineers to work either in the time domain or the frequency domain as follows:

- In the frequency domain: Fourier transforms;
- In the time domain: Laplace transforms.

It is beyond the scope of this work to pursue such analyses further.

Increasingly, bandwidth-hungry applications go from left to right across this entire chart, starting with voice and low-bandwidth internet connections and proceeding through to high-level multiplexed applications. This is summarized in Figure 2.5 where the bandwidth demand is plotted to a base of the increasingly demanding applications. Voice, generally phones, is the least demanding, and, in contrast, entertainment TV (see Chapter 5, particularly high-definition TV (HDTV)) is the most demanding of bandwidth. It should be noted that this chart is generated on a per channel basis—that is, individual single channels are considered here.

Signals associated with all these applications are frequently transmitted by wireless or optical (including fiber-optic) means, and, prior to this, high-level multiplexing is applied in a manner similar to that shown in Figure 2.2. The final bit rates, often hundreds of

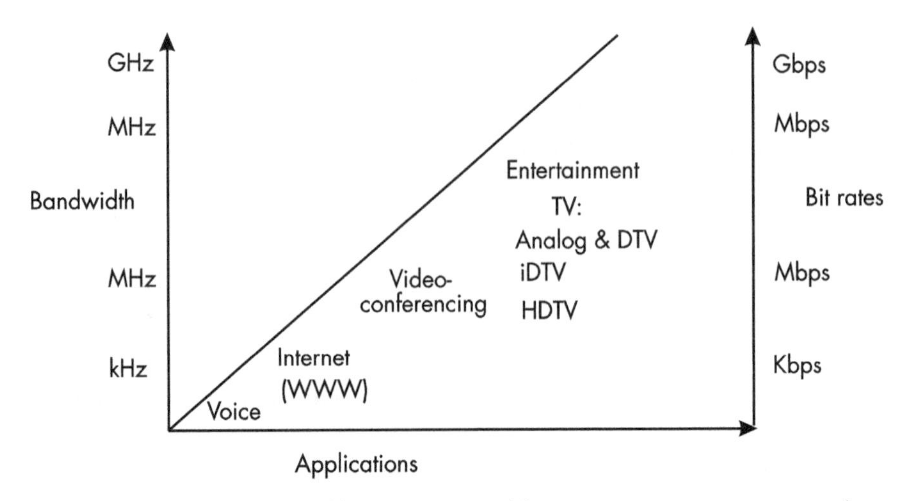

Figure 2.5 Bandwidths and bit rates required for various communications applications (examples of videoconferencing include Microsoft Teams and Zoom). *From:* [4].

megabits per second or gigabits per second, require substantial bandwidths, and the total bandwidth available for such signals is limited in all instances.

What is often disregarded, at least until an excessively late point in considerations and plans for new systems, is that the available bandwidth in radio (microwave/millimeter-wave) and optical systems depends markedly upon the region of the electromagnetic spectrum in which the system is desired to operate.

In general terms, this more or less means "you can't fit a quart into a pint pot."

2.2 CURRENT AND FUTURE BANDWIDTH RESOURCES

2.2.1 Bandwidth and Bit Rate

There is often confusion between bandwidth and bit rate and the true meaning of these terms. The foregoing discussions should clarify just what bandwidth and what bit rate amount should be, as well as the fact that as bit rate demands increase so indeed the required bandwidth for transmission also increases.

So here is an absolutely key question: If you double the bit rate demanded, do you also simply double the bandwidth required for transmission? Unfortunately, the answer is negative. Things are just not that simple; they hardly ever are. This is because, aside from bandwidth and bit rate, there are two further important considerations:

1. The modulation scheme chosen;
2. The relative amount of noise power in the channel.

With digital communications systems, several modulation schemes are available, for example:

- Quadrature amplitude modulation (64QAM, 256QAM, nQAM);
- Quadrature phase-shift keying (QPSK).

Even with constant noise, in all instances, the bandwidth required will always differ according to the choice of modulation scheme. In many instances, the selection of modulation scheme is at least partly contingent upon its bandwidth efficiency, that is, the maximum bit rate practically achievable using a given bandwidth.

Unfortunately, it is not only our wanted signals that are present throughout transmission. Instead, we have to contend with pervasive unwanted disturbances, generically called noise, that adversely influence all electrical and optical systems. The average power arising from this noise can often itself be broadband, therefore affecting all signals across the spectrum. A major objective in communications systems design and implementation is to be able to distinguish the desired signals from the background noise and then receive the messages accurately. There was an early discovery that formed a very important guide, long before digital communications were even possible.

In the 1940s, a brilliant communications theorist named Claude Shannon (working in conjunction with Hartley) discovered that, even for basic digital binary modulation, bit rates could never simply increase proportionately with bandwidth. The effects of the background noise in communication channels also clearly impact the bandwidth required to transport a given signal at a known bit rate. Shannon's classic formula is:

$$C = B \cdot \log_2 \left\{ 1 + \left({}^{S}\!/\!_{N} \right) \right\} \tag{2.1}$$

where

C is the channel capacity in bits/s (upper bound);

B is the channel bandwidth in hertz;

S is the average received signal power in watts or v^2;

N is the average white Gaussian noise power (AWGNP) plus interference in watts or v^2.

Notice that S/N is necessarily a dimensionless ratio.

Also note that Shannon's formula is logarithmic (base 2) in terms of the channel signal-to-noise ratio. This law tells us the maximum possible bit rate that may be transferred in a given channel having bandwidth B (Hz) in the presence of a specific amount of noise power. Base 2 occurs due to the binary nature of digital transmission (1 or 0). Shannon's result gives us the maximum limit bit rate. Using any of the known modulation schemes and the best transmission channel that we can find, it would not be possible to exceed this limiting bit rate.

It is necessary to appreciate that, within a specific radio or optical link, the total bandwidth actually contains typically several thousands of digital channels, each only occupying a few kilohertz of bandwidth. However, because of the random nature of the noise, increasing the bandwidth may do little for the signal and yet increase the noise considerably, making communications more difficult with most systems. Engineers talk in terms of the ratio between the signal power and the average noise power in the channels.

It is useful to provide a numerical example and, for this purpose, an online link [1] is used as follows (warning: S and N have to be input as linear numbers, not decibels):

Calculate the channel capacity where the input data points are: Bandwidth = 400 MHz, signal power = 5 dBm, and noise power = 2 dBm. Initially, both S and N are converted to linear quantities (i.e., watts or milliwatts). The resulting linear ratio is then substituted into (2.1). Processing using this equation leads to:

Channel capacity = 193.82 Mbps.

Clearly, the channel capacity suffers (decreases) because of the noise in the channel. As a consequence of this noise, the capacity is substantially less than half the numerical bandwidth.

In order to successfully transport signals, in most instances, the signal power has (at least on average) to exceed the average noise power. With some innovative systems, the signal can be literally buried within the noise and yet accurate reception remains possible. This approach is referred to as spread-spectrum and very wide bandwidths are used [2]. The concept of bandwidth efficiency, referred to above, is extremely important. Additionally, the choice of carrier frequency is also highly significant. Further details including those concerning digital modulation options can be found in [2]. Various levels of QAM are frequently encountered in both radio and cabled systems and these are summarized in Section 2.2.2.

2.2.2 Bandwidth Efficiency or Spectral Efficiency

The bandwidth efficiencies associated with a chosen modulation scheme are very important and the efficiencies are quoted in Table 2.1 for a selected range of modulation schemes.

Most of the data in Table 2.1 is available in [2, Table 14.2, p. 257]. Further information was obtained by surfing the internet.

There is an important caveat to the data shown in Table 2.1. While it looks attractive to simply go on increasing the modulation level (64 QAM, 256 QAM, 1,024 QAM), the issue is that, because of the ever-smaller amplitude steps, there is an accompanying increase in the generation of errors. This leads to an increasing systemic probability of error (bit error rate (BER)), and it may be wise (if possible) to try keeping the modulation scheme to relatively modest levels such as 256 QAM.

As well as amplitude steps, another problematic factor is phase steps, although channel coding is usually applied in order to minimize the effects of small phase steps.

All the above details apply regardless of the technology considered, or, in most cases, regardless of the applications: radio, optical

Table 2.1
Bandwidth Efficiencies

Type of Modulation	Bandwidth Efficiency (bps/Hz)
QPSK	2
64 QAM	6
256 QAM	8
1,024 QAM	10

(photonic), terrestrial, and space. Technologies appropriate for digital and RF chips were described in Chapter 1 and the systems as such are discussed in Chapters 4, 7, 8, and 9. Adopting a similar approach, we now present some technologies that apply to optical photonic communication links: free-space or fiber.

2.3 TECHNOLOGIES FOR OPTICAL SYSTEMS

2.3.1 Some General Aspects of Optical Systems

As the twenty-first century progresses, so applications of photonic technologies into communications continue expanding ever more strongly. In general, these comprise fiber optics and free-space optics (FSO).

At a systems level, fiber optics is described in Chapter 3, while a specific application of FSO is described in Chapter 4. In this section, we concentrate on introducing a selection of key components required for fiber-optic and also FSO links.

In the earlier sections of this chapter, we focused strongly on bandwidth, a much-prized feature of any communications system (i.e., applying to optical systems as well as radio). However, the first key feature to understand is that, with optical communications, the available bandwidth is relatively massive. Optical fiber offers a total available bandwidth of 25 THz centered on the 1,550-nm wavelength, which is the middle of optical C-band. Bear in mind that 1 THz equals 1 million times 1 million Hz, or 1,000 GHz. Millimeter-waves are in the tens of gigahertz ranges (see Chapters 1 and 8), and 6G requires subterahertz technology (Chapter 10). Optical communications operate at carrier frequencies such as 200 THz (i.e., wavelengths in the region of thousands of nanometers).

Theoretically, for FSO, vast swaths of bandwidth are available. The total available bandwidth for fiber optics is also very large, but is limited by the fiber itself.

2.3.2 Masers, Lasers, Photodetectors, and Optical Amplifiers

The mid-twentieth century turned out to be one of the most exciting periods for highly fruitful technological discoveries, driven by clear needs that were important for many reasons. Examples closely related to microwaves and also photonics include two very important inventions, both of which include "SER" in their acronyms: the maser and the laser.

The first of these devices, the maser, stands for microwave amplifier (based on the) stimulated emission of radiation, and this was invented in 1953 by Townes, Gordon, and Zeiger at Columbia University in the United States. The main driving force behind the maser comprised the need for a very low noise amplifier (LNA) that would be capable of amplifying the extremely low-level, satellite-originating signals being received by the ground stations around the early 1960s (Early Bird). In this respect, masers were remarkably successful, but quite soon much more compact, convenient, and often wideband solid-state LNAs completely took over this role, and indeed this is still the case today (GaAs LNAs, GaN LNAs).

The second device mentioned above, the laser, stands for light amplifier (based on the) stimulated emission of radiation, and the first-ever laser was demonstrated around 7 years after the invention of the maser. Although the "a" in laser means amplifier, all these devices are really oscillators because they principally generate a beam of single-frequency light. Theodore Maiman invented the laser, and on May 16, 1960, he demonstrated the first such device at Hughes Research Lab in California. This device comprised a fairly large and cumbersome lab-based structure, but it paved the way for extremely useful devices designed for applications ranging from kilowatt-power precision industrial cutting machines to relatively small semiconductor lasers that form the main light transmitting element in optical communication systems. For obvious reasons, we concentrate on these small semiconductor lasers.

2.3.3 Semiconductor Lasers

For many years, light-emitting diodes (LEDs) served as the transmitting devices for FSO and in earlier times even for basic fiber optics. During the 1970s, what became known as distributed-feedback lasers (DFBs) soon took the optical communications world by storm and, along with VCSELs, DFBs remain of great importance today. A schematic outline of a DFB laser is shown in Figure 2.6.

Fundamentally a diode structure (P-type through to N-type), every laser requires a DC supply of at least several volts, connected

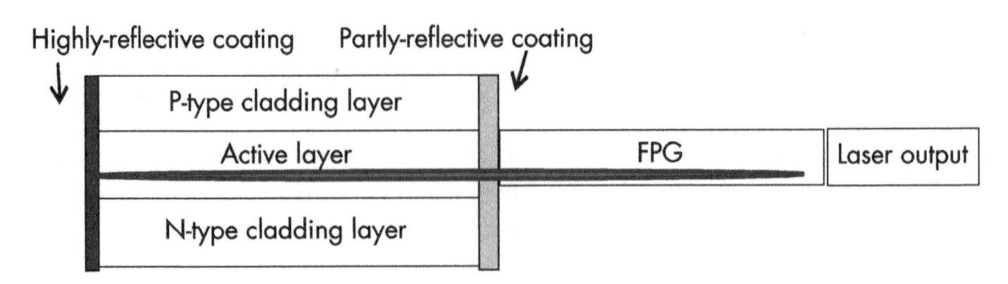

Figure 2.6 Schematic outline of a semiconductor DFB laser (the semiconductor materials typically comprise various complex alloys involving GaAs).

across this structure, usually termed the DC bias. Properly biased, this device can generate a laser output amounting to typically some tens or hundreds of milliwatts. This output always possesses very high spectral purity (i.e., a dominating single wavelength typically 1,552 nm (1.552 μm) accompanied by very low-level unwanted side wavelengths).

In Figure 2.6, Fabry-Perot grating (FPG), also known as a Bragg grating, comprises a multiridged structure through which the rapidly growing light beam repeatedly passes. This is illustrated by the blue region within the active layer and the FPG. The beam repeatedly passes through the active layer, which is where the power increases because the beam power is being extracted from the diodes' DC supply. The highly reflective coating shown at the left end (Figure 2.6) ensures that the beam is repeatedly reflected back through the entire structure. The partly reflecting coating allows most of the beam to be continually growing and to pass through the FPG and emerge as the laser output.

There are also quantum cascade lasers (QCLs) that indicate potential for future terahertz-modulated communications systems such as 6G [personal communication with Professor Ian Robinson, Electrical Engineering, University of Leeds, United Kingdom, 2023].

Vertical-cavity-emitting semiconductor lasers (VCSELs) are also important and a three-dimensional (3D) schematic diagram of this type of laser is shown in Figure 2.7 [3].

A good description of VCSELs is available in Geoff Varrall's book [3].

Semiconductor lasers are also occasionally tuned using external cavities, but further details are beyond the scope of this book.

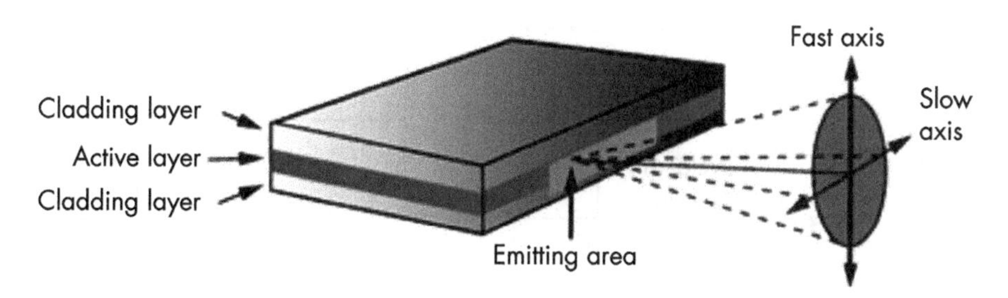

Figure 2.7 Elliptical beam pattern from a VCSEL. (*From:* [3]. Reprinted with permission.)

2.3.4 Photodetectors

The first stage at the receiving end of any optical link always begins with some form of photodetector. This is usually again a diode-type component; it can be a fairly simple photodetector diode for relatively short-range links carrying channels with maximum speeds in the megabits per second class. For faster signals, an avalanche photodetector is required.

Like DFB lasers, avalanche photodetectors have been available for many decades and are regularly implemented today. Following this first detecting stage, it is necessary to amplify the electrical signal, and this is almost accomplished by what is known as a transimpedance amplifier. Additional passive circuitry is required in order to match the output impedance of the transimpedance amplifier, which is always far removed from that of the usually 50-ohm microwave technology. The microwave electronics is required for further signal processing in the receiver signal processing chain; see Chapter 1.

2.3.5 Optical Amplifiers

Erbium-doped fiber amplifiers (EDFAs) are available for optical signal amplification within this band, although none can cover more than about one-eighth of the entire band. Erbium is classified as a rare Earth element, although it is actually quite abundant. EDFAs typically provide around 20 dB to 25 dB of gain over 3-THz bandwidths, and the typical (block) arrangement of such an amplifier is shown in Figure 2.8.

In common with any amplifier, there is a signal input port and a signal output port. The source of optical power, from which the signal

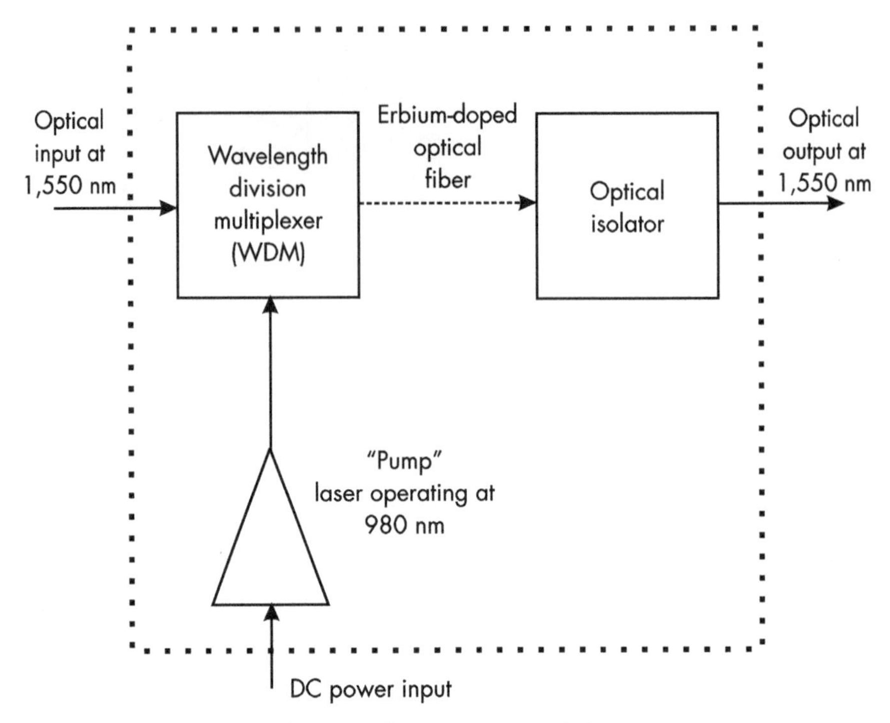

Figure 2.8 Functional schematic of an EDFA. *From:* [4].

gain is derived, is a semiconductor "pump" laser that usually operates at 980 nm and the output from this laser feeds into a wavelength division multiplexer (WDM). The combined input signal and the WDM power from the laser are fed through a length of erbium-doped fiber in which the amplification actually takes place. An output optical isolator ensures that negligible amplified and reflected output power reenters the amplifier (this could cause oscillations, or worse). WDM and DWDM are described in Chapter 3.

EDFAs are now vital elements in long-distance optical systems and are becoming increasingly important for very-high-bit rate moderate-range networks. Manufacturers include Cycle, CSRAYZER Optical Technology, Lumibird, Menlo Systems, MPB Communications, RPMB Lasers, and Thorlabs.

Maximum power gain is typically 20–25 dB across the optical C-band (1,530–1,565 nm) and output signal power usually amounts to around 1W to 2W.

All-semiconductor optical amplifiers (SOAs) are also available, but EDFAs are the most important class of products used.

The foregoing descriptions lead us to the study of broadband cabled and free-space optical networks, which is the subject of Chapter 3.

References

[1] www.satcomresources.com/shannon-hartley-channel-capacity-calculator.

[2] Edwards, T., *Technologies for RF Systems,* Norwood, MA: Artech House, 2018.

[3] Varrall, G., *5G and Satellite RF and Optical Integration,* Norwood, MA: Artech House, 2023.

[4] Edwards, T., *Gigahertz and Terahertz Technologies for Broadband Communications,* 1st ed., Norwood, MA: Artech House, 2000.

3

BROADBAND CABLED AND FREE-SPACE OPTICAL SYSTEMS

3.1 INTRODUCTION

In this chapter, considering broadband optical connections, we present information on both cabled and FSO systems. More specifically, CATV and SMATV are discussed in Chapter 6. Satellite-based systems are covered in Chapters 4, 7, and 9, and terrestrial and stratospheric radio networks are the subjects of Chapter 8.

3.2 NEW GENERATIONS OF FIBER-OPTIC CABLED OPTICAL SYSTEMS

3.2.1 The Evolution of Fiber-Optic Technology

In 1966, a research group headed by two now-famous researchers named Kao and Hockham, working at what was known as Nortel's U.K. research center, published a landmark theoretical article on the transmission of optical signals in glass strands. The article predicted that such signals could be successfully transmitted along silica or glass waveguides, but there was one somewhat serious problem: the attenuation would probably be in the region of some tens of decibels per kilometer. At the time, this provided further fuel for the copper circular waveguide enthusiasts with their relatively low 3 dB per km.

Just 4 years later in the United States, Kapron and Keck demonstrated losses of only 20 dB per km in practical optical fibers, and, in 1976, the Rediffusion Company (London, U.K.) introduced an optical fiber cable connecting cable TV subscribers as the first commercial application (Chapter 6 covers CATV and SMATV). Six years later in 1982, the first single-mode fiber cables emerged with losses down to an incredible 0.2 dB per km—an order of magnitude below the best copper waveguide results. This figure is very close to the absolutely lowest limit achievable for such cables, and even today 0.2 dB per km is commercially state of the art.

Worldwide, the markets for fiber-optic equipment of all forms—cables, connectors, transceivers, amplifiers, test equipment, and so forth—continue to grow steadily and this is expected to continue accelerating over the years ahead. The chart shown in Figure 3.1 indicates this global growth regarding networking equipment.

From Figure 3.1, it can be seen that the markets approach $14 billion a quarter-way into the twenty-first century and are forecasted to exceed $25 billion by 2030.

There are two principal classes of optical fiber: multimode and single-mode. To the uninitiated, it may at first seem that multimode cabling must surely be superior technically to single-mode, but the opposite is the case. In multimode optical fibers, many modes or rays of transmission are set up and these travel down the fibers in zigzag

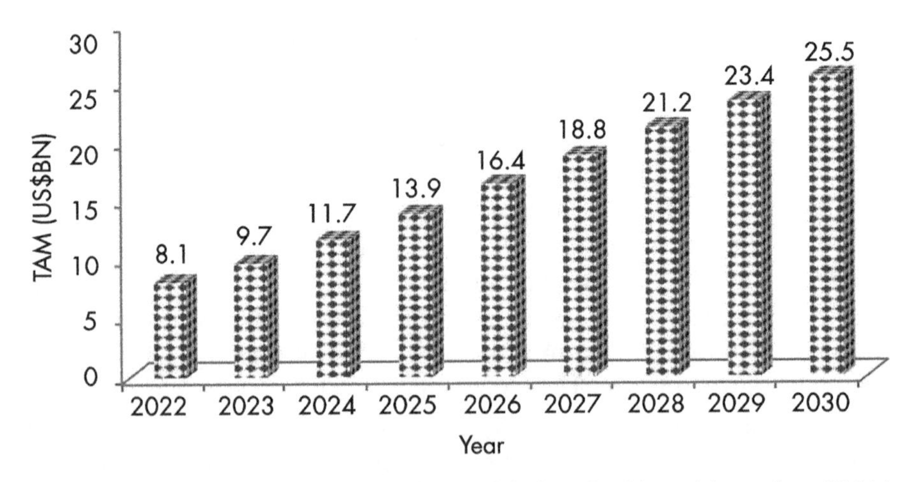

Figure 3.1 Current and forecasted future global total addressable markets (TAMs) for optical networking equipment in billions of dollars (data generated by the author).

fashion (a convenient fact that led to fundamentally sound analysis in the 1970s). Also, the optimum wavelengths for transmissions are typically 630 nm, 820 nm, or 1,300 nm. Multimode fibers exhibit considerably more attenuation at these wavelengths than at the 1,550-nm optimum applying to single-mode.

At best, we are talking several decibels compared with the 0.2-dB theoretical minimum. It is not only this attenuation feature that leads to the reach distance limitations with multimode fibers, because another important restriction known as dispersion represents a serious issue.

Unfortunately, all fibers of all types, in fact, all transmission media, exhibit dispersion. The zigzag trajectories referred to above are a major source of dispersion in multimode fibers, and this is characterized by different times of arrival for different rays as these travel along the length of cable. As a result, digital signals arrive out-of-sync at the receiver—some are delayed compared with others. Beyond a definite limit (Figure 3.2), this dispersion produces what is called intersymbol interference (ISI) at a level that is intolerable for the system's operation. Controlling this ISI is a major goal for the system designers.

However, there is a highly significant advantage with multimode fiber cables and this is the relative ease and economy of interconnections, including the demountable components. This applies because the dimensions of the multimode cables are inherently much larger than those for single-mode: tens to hundreds of microns compared with just several microns.

A typical multimode fiber structure will have a 50- or 62.5-micron core (the smallest internal diameter) and a 125- or 150-micron outer diameter. In comparison, a single-mode fiber has a 5 to 8-micron core and a 50-micron outer dimension. Handling the multimode cable is much easier and more reliable than is the case with single-mode. This fact continues to place emphasis upon multimode technology wherever possible, and the power penalty for maintaining ISI at tolerable levels is shown in Figure 3.2, where the parameter is the fiber bandwidth • distance product (often simply called bandwidth) in megahertz or gigahertz per kilometer. The power penalty is a measure of the amount of additional transmitter power required to maintain a tolerable ISI.

At 500 MHz per km, the best multimode fiber shown in Figure 3.2 has the characteristic that, even over half a kilometer reach

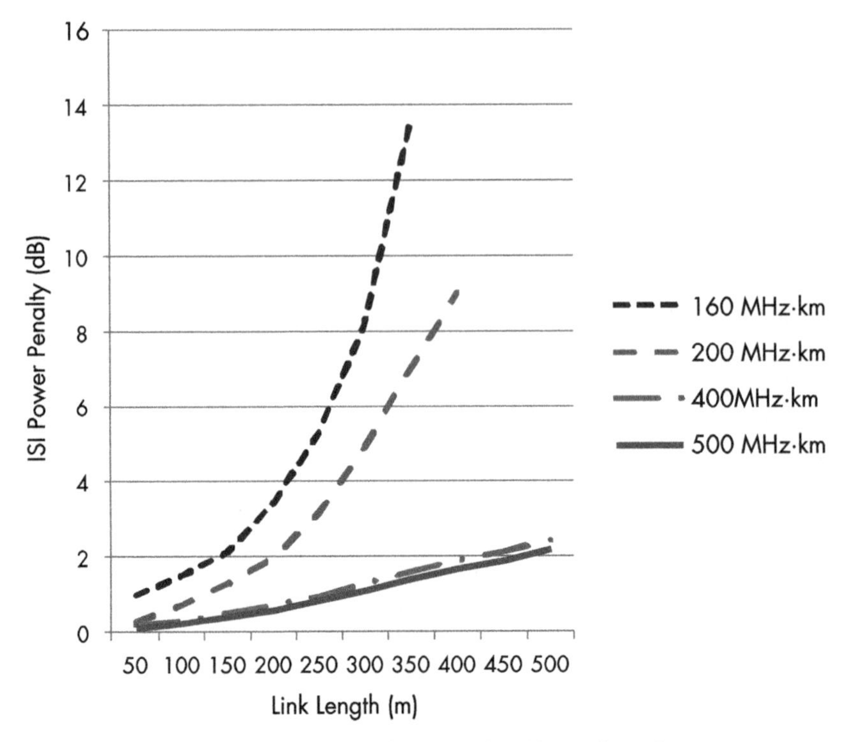

Figure 3.2 ISI penalties for different fiber bandwidths (multimode).

distance, only about 2-dB power penalty is incurred. The following general maxim can also be used for multimode cables (applying at least through the mid-2020s: Using a multimode fiber carrying a signal with a bit rate of 10 Gbps, the maximum reach is about 550m (before the ISI penalty becomes intolerable).

In summary, multimode fibers are mainly used in applications such as low-cost CATV, LANs, and the local loop. In these kinds of applications, an important practical concept involves the optimum combination of coaxial (copper) and multimode fiber cabling known as hybrid fiber-coax (HFC).

Single-mode fibers have principally been used in long-distance, often sub-oceanic, national, and international trunk cabled applications and also increasingly even in much shorter routes (see Chapter 6).

For long-distance communications, a fiber technology known as soliton transmission is often adopted today. With this technology, special optical pulse shapes are transmitted, enabling tens of kilometers to be spanned without intervening amplification and regeneration.

The SDH and synchronous optical network (SONET) introduced in Chapter 1 are of such importance that they are considered next in more detail.

3.2.2 Synchronous Digital Hierarchy, Synchronous Optical Network, and Fiber

As mentioned in Chapter 1, digital signal framing and multiplexing are of major importance. This leads, in turn, to basic standard requirements for bit rates (or data rates) for transmission at various levels through the networks.

Historically, the bit rates have differed in various countries and world regions as a result of earlier evolution that was based largely upon national considerations, including standards. Countries such as Japan, the United States, the United Kingdom, and European nations all had somewhat different and mutually incompatible digital hierarchies. In North America, for example, the digital signal (DS) hierarchy series prevailed although this involved noninteger multiples of the basic PCM voice channel rate (64 Kbps) at the first two levels: DS-1 and DS-2. Figure 2.2 of Chapter 2 indicates the basic SDH multiplexing process.

Today, globalism is the watchword in many major respects and having access to internet-based communications is regarded as being almost as important a service as water and electricity. This means it is vital to be able to intercommunicate without having to translate from one digital hierarchy to another different system. Apart from hardware and software requirements, this also risks substantial timing errors in the digital signals. As a result, the International Telecommunication Union (ITU) agreed upon what is now universally known as the SDH. Apart from the first optical level, known as OC-1, all the SDH rates are precisely equal to corresponding optical rates as shown in Table 3.1. Timing errors are kept to a minimum by adopting atomic frequency standards and Global Positioning System (GPS).

In Table 3.1, OC means optical carrier, SONET refers to synchronous optical network, STM means synchronous transport module (i.e., the information frame structure), and STS means synchronous transport signal, which is a term used in conjunction with SONET. Although the primary focus here is upon optical networks, it is

Table 3.1
SDH and SONET Digital Signal Bit Rates

ITU-T: SDH (STM-m)	SONET (STS-n)	Optical Carrier (OC-n)	Bit Rates (Mbps, Gbps)
	STS-1	OC-1	51.84 Mbps
STM-1	STS-3	OC-3	155.52 Mbps*
STM-4	STS-12	OC-12	622.08 Mbps
STM-16	STS-48	OC-48	2.48832 Gbps
STM-64	STS-192	OC-192	9.95328 Gbps
STM-256	STS-768	OC-768	39.81312 Gbps
STM-1,024	STS-3072	OC-3072	159.25248 Gbps

After: Nellist and Gilbert [1].
Note that each new level is specified four times its predecessor.
*This is the gateway level, that is, the principal interconnection rate for connected networks.

important to understand that all the SDH (and therefore also STM-N) references apply equally to FSO networks as well as microwave and millimeter-wave radio systems including satellite.

Information for Table 3.1 was largely taken from Nellist and Gilbert's book referred to above (their Table 1.1), as well as my first edition of this book, and these sources are acknowledged.

Internet data (especially 5G) is increasingly taking the lion's share of the traffic. Note that the STM-m bit rates referred to in Table 3.1 increase progressively by factors of 4 (binary 2^2). Signals at OC-768 level are increasingly required (almost 40 Gbps) and OC-3072 is now being frequently encountered. A block diagram illustrating how the SDH signals are developed is provided as Figure 3.3. The aim is to group several signals to form an STM-16 output information stream at 2.488 Gbps. A wide variety of input signals is shown, ranging from the "D1" (or DS-1) 1.544-Mbps signal, through the 2.048-Mbps European E1 level, followed by the 6.312 DS-2 and 44.736-Mbps DS-3 instances, and finally the relatively high bit rate DS-4 at 139.264 Mbps. The bit rate incompatibility between standards is clearly a serious issue.

The components C-n are data containers, AN are aligning units, MUX are digital multiplexers, and AUG is the administrative unit group feeding the final output. Note how the lowest levels (i.e., slowest) input signals require the greatest amount of aligning and multiplexing before reaching the final output multiplexer. In contrast, the

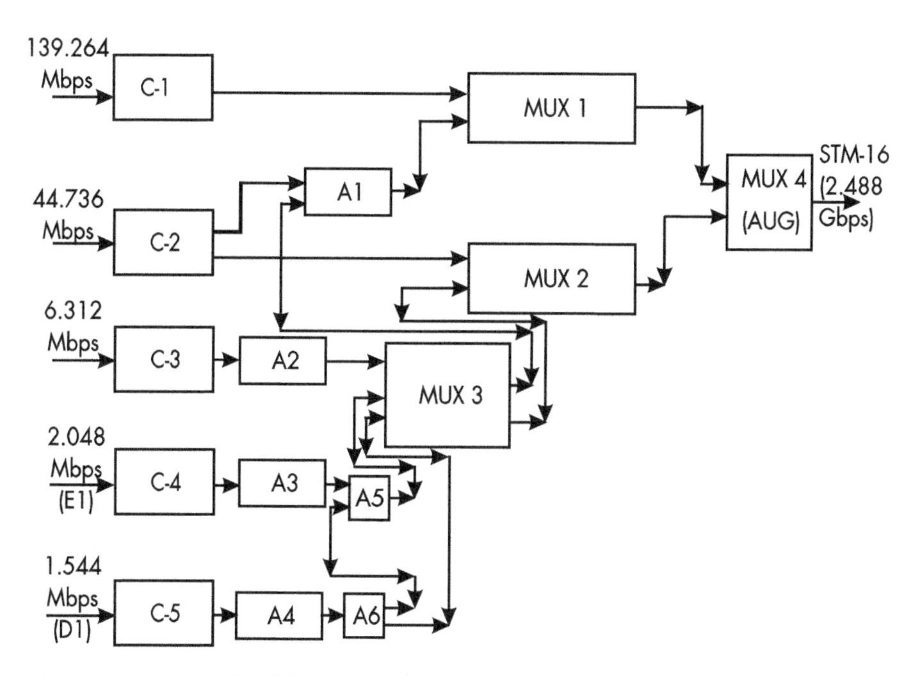

Figure 3.3 Example of the SDH multiplexing process (to STM-16 level). *From:* [5].

fastest input signal in this example (139.264 Mbps) can be passed directly into the final MUX.

The signal processing technology used is generally conventional silicon electronics with RF CMOS for the higher speeds at least (see Chapter 1 for details concerning the choice of technology).

Practical, commercially available technologies use an internal laser-based transmitter typically operating at 1,530–1,565 nm with reach distances of many kilometers necessarily adopting single-mode fiber cabling.

3.3 GIGABIT DATA NETWORK EVOLUTION

3.3.1 Local Area Networks and Fiber Distributed Data Interface

Local area networks (LANs) have established a secure niche in data communications. Early LANs supported backbone (bus) bit rates only as high as about 10 Mbps, but fiber-based networks soon pushed this up to 100 Mbps (and above), using the fiber distributed data interface (FDDI) and fast Ethernet.

FDDI does not have to use fiber as the bus physical layer, as the name refers to a specific transmission protocol. Where link lengths are sufficiently short (generally much less than 550m) and interference is not a serious issue, copper conductors may be used. This is akin to the comment made above regarding hybrid fiber-coax (HFC) and it may be acceptable to adopt copper distributed data interface (CDDI), where this all-electronic technology is adopted.

Although not strictly a broadband data network by current standards, the FDDI LAN backbone is briefly considered here because it paved the way for much faster standards.

This LAN protocol normally employs 4B/5B coding that allows for a maximum line transmission rate of almost 125 Mbps. This 4B/5B coding operates by taking 4 bits of the signal in blocks and the approach is considerably more spectrally efficient than many other methods. Using fiber as the transmission medium, total path lengths up to 200 km are possible, that is, 100 km of single fiber because a dual-fiber configuration is implemented. With this basic protocol, 500 nodes or stations can be supported, but parameter values may be increased to allow for much larger networks.

An overall schematic block diagram of FDDI (simplified by identifying only six stations) is shown in Figure 3.4. In this example, each active node (station) regenerates the required signals. In practice, many more nodes are implemented and one will be the server station. Some stations just operate in bypass mode, not playing a part in the LAN operation, and this is illustrated with station number 3 in Figure 3.4.

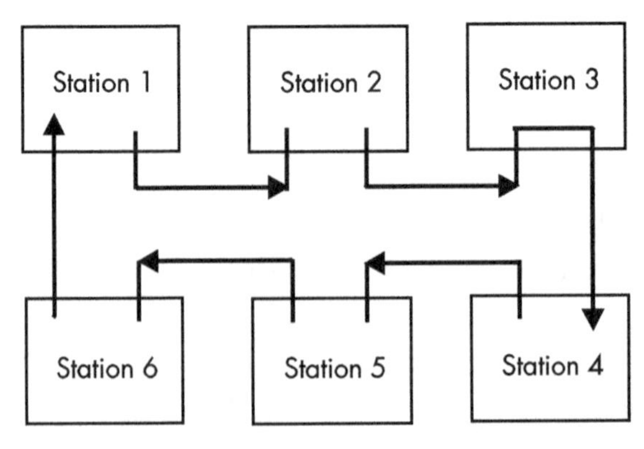

Figure 3.4 Schematic block diagram of FDDI (simplified). *From:* [5].

FDDI basically operates on the token ring principle in which a unique control signal known as the token circulates on the transmission bus medium. Only the active stations are primed to react to the presence of the token which is captured and removed from the ring before that station transmits its message signal. At this time, the station issues a new token. Some stations can actually comprise complete sub-LANs in their own right. For fiber-based FDDI LAN backbones, multimode cables are mostly implemented, with wavelengths of either 820 nm or 1,300 nm.

As mentioned earlier, LANs began with exclusively copper-based transmission layers, and many used the Ethernet protocol that was originally developed by Xerox Corporation many decades ago. The fundamental Ethernet topology is shown in Figure 3.5.

Only six stations are shown connecting to the bus bidirectionally and, in practice, these types of LANs cannot support large numbers of stations where copper transmission applies.

As unit prices for gigabit-level fiber-optic components have continued to fall over time, so systems integrators have seized the opportunity to introduce this technology into LANs to a much greater extent than ever before. The dramatic growth of all forms of communications, particularly data, continue to drive onwards all appropriate technologies.

The simplicity of the Ethernet topology with its ready extendibility led to the development of what has become known as Gigabit Ethernet, which means the basic Ethernet topology implemented with a fiber transmission bus operating at tens or hundreds of gigabits per second or even higher.

The latest developments in fiber networks blend together the fundamental approaches and technologies associated with ATM, SONET, LANs, and DWDM. This leads to the following very important aspects of twenty-first-century fiber-optic networking: DWDM and passive optical networks (PONs).

3.3.2 DWDM

FDM, applicable to electrical signals, was described in Chapter 2, where Figure 2.1 illustrated the principle. In the optical world, wavelengths are conventionally referred to rather than frequencies, but the

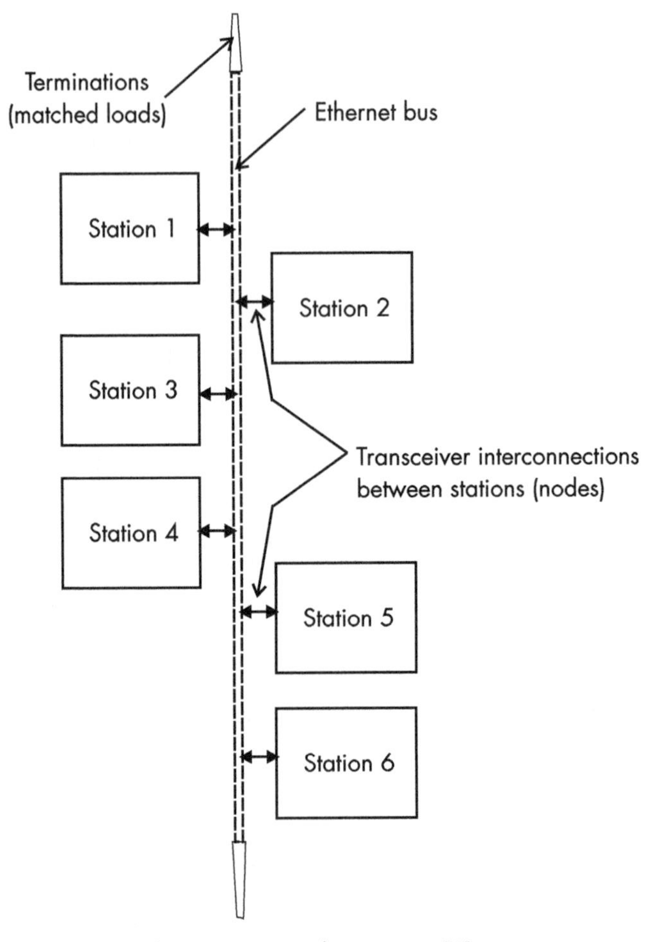

Figure 3.5 The basic Ethernet LAN topology. *From:* [5].

basic principle is the same, that is, information channels are segregated into different wavelength ranges.

By analogy with the electrical case, this process is called WDM for fiber-optic transmission. Within each wavelength range, the optical signals suffer a small amount of attenuation, whereas beyond the allocated channels, the attenuation rapidly rises to typically several tens of decibels. WDM is mainly applied to single-mode systems, but it can also be used with multimode networks. The principle is shown in Figure 3.6.

Light signals from laser sources each operating at slightly different wavelengths are fed in parallel into the wavelength multiplexer,

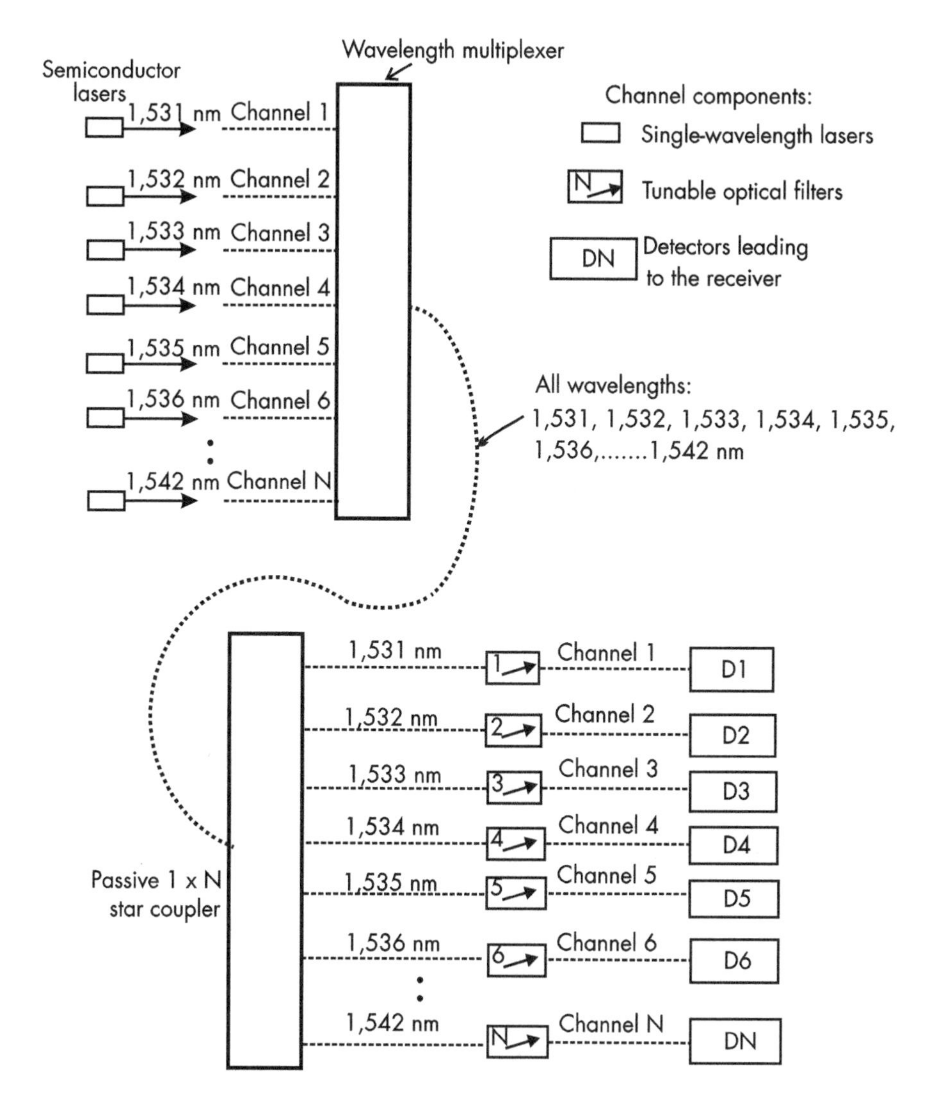

Figure 3.6 WDM. *From:* [5].

which then outputs a single continuous optical signal made up from all *N* of the input wavelengths. The fiber cable carrying this combined signal must have sufficient bandwidth to be able to cope with what is often a very large capacity. However, the bandwidth of single-mode fiber at 1,550 nm is extremely large (approximately 25 THz), so that up to several thousand such sources can be applied simultaneously. At or close to the receiver, a passive star coupler takes this combined signal as its input and delivers an attenuated version into N output ports, where N is the number of original wavelengths. Following each output port, a tunable filter selects one of the required wavelengths (rejecting all others) and passes this signal to the receiving detector. The final signal is subsequently delivered to the user after conversion to electrical form. A passive star coupler is an optical component with one input port and at least as many output ports as there are wavelengths. Wavelength tuning filters are made available by exploiting the wavelength selective properties of passive optical devices such as diffraction gratings, electro-optic filters, and acoustic-optic filters. The actual tuning can be achieved mechanically, for example, by rotating a diffraction grating, or (much faster and more effectively) electronically by attaching the filter to a semiconductor chip and varying the applied voltage.

Most WDM components use miniaturized optical structures based upon diffraction grating or Bragg principles in which light beams emitted by the different lasers become focused into a single stream that is coupled into the output fiber. All WDM components are therefore passive because no active semiconductor or other oscillating, amplifying, or switching function is involved.

WDM first became available during the 1980s when Deutsche Telecom, for example, introduced the technology into their BIGFON metropolitan networks. Ten-channel WDMs (state-of-the-art at the time) were implemented with 36-nm channel spacings and 20-nm signal pass bandwidths. Manufacturers included AEG (now a part of Daimler-Benz), Amphenol, and Corning-France.

By the turn of the millennium, WDM technology had not only become much more sophisticated; it had also become vastly denser and had entered major growth markets. The era of WDM as such gave way to what is now being called DWDM.

With this technology the channel count raced towards the 130-plus realm, unthinkable by 1980s standards. In 1999, for example, Lucent Technologies won its first order for an 80-channel system, and Pirelli offered a system that was scalable to 128 channels. Nowadays DWDM systems with 1,000-plus capabilities are commonly encountered.

For most systems integrators, it is the implementation of some degree of future-proofing that is particularly significant because the economics of repurchasing and reinstalling entirely new systems places substantial financial burdens on the companies. Almost any technology that is affordable now and yet offers the real prospect of being cost-effectively upgraded in the future is likely to win important votes at the boardroom level. DWDM manufacturers include Ciena, Fujitsu Telecoms, and Lucent Technologies. Ciena/Juniper, for example, offers an OC-48 router with a capacity of 40 Gbps transmitting over a distance of 12,000 km.

Careful tuning of DWDM networks is vital and self-tuning is increasingly implemented [2]. Previously, retuning was performed manually, using high-reach "cherry-picking" equipment, but self-tuning a DWDM transceiver means automatically scanning and selecting the desired new wavelength.

3.3.3 Passive Optical Networks

3.3.3.1 The Passive Optical Network Concept

Until the 1980s, almost all fiber-optic networks required full-on transceivers at every node-multiplied for every network. This topology demands repeated nodal resynchronization as well as:

- Incurring the substantial capital costs of acquiring all the transceivers;
- Coping with the considerable operating costs and adverse reliability issues;
- Coping with the amount of energy wastage.

All these factors represent important considerations, and this led to the concept and early introduction of PONs in which there is only one active node containing the transceiver. The overall concept is illustrated in Figure 3.7.

A PON is a cabling system that uses optical fibers and optical splitters to deliver services to multiple access points. A PON system can be fiber-to-the-curb (FTTC), fiber-to-the-building (FTTB), or fiber-to-the-home (FTTH). Such a system comprises optical line terminations (OLTs) at the communication provider's end and a number of optical network units (ONUs) at the user's end. The term "passive"

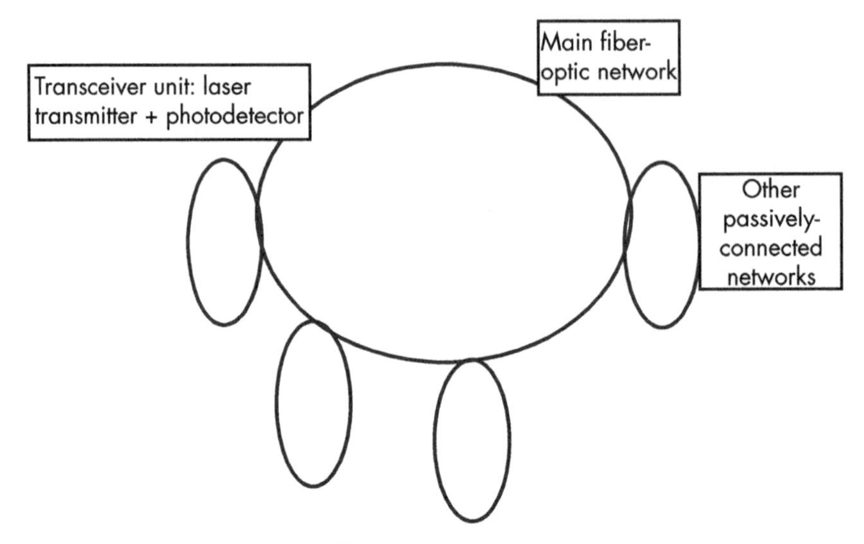

Figure 3.7 Schematic outline of a PON.

simply means that there are no power requirements (except the single transceiver) while the network is up and running.

One important feature is that PON port costs are substantially lower than WDM port costs, making PON particularly attractive for 5G networks.

Ultra-broadband and coexistence with existing technologies are the common requirements from network operations to support PON evolution. Operators across the globe are seeking to increase revenue by developing bandwidth-consuming services. The service includes HDTV, which requires about 20 Mbps per channel. In the 2020s, new business models, such as home video editing, online gaming, interactive e-learning, remote medical services, and next-generation 3D TV all serve to considerably increase the bandwidth demand. To meet these demands of increasing bandwidth the high transmission speed of 5G networks coupled with secure and reliable network operation between the end points are fueling the growth of the market. It also interfaces well with the coming 6G (see Chapter 10).

3.3.3.2 Passive Optical Network and 5G

The increasing demand for 5G technology to compensate for the rising number of mobile internet users is accelerating the growth of the

PON market and will almost certainly continue to progress this way. Some key benefits of PON in the 5G network are:

- It reduces the cost of deploying a fiber access network by reducing the number of network termination ports.
- The fiber length is typically 20 km for passive systems, which enables a significant level of node consolidation.
- It eliminates recurring costs associated with the fabric of Ethernet switches in the network.
- Network maintenance is easier and less expensive than otherwise.
- Amplifiers in the systems enable PONs to extend the reach to beyond 100 km.
- PON infrastructures tend to be generally smaller than would be the case with active infrastructures.

5G networks require significant upgrades to optical networking in order to support already wide bandwidths as well as the various bandwidth-intensive applications. PON, with its point-to-multipoint architecture, plays a key role in the massive 5G rollout and cell densification. Therefore, the implementation of next-generation PON is already being intensified in this 5G era and the selection of next-generation PON is a trade-off between technology and cost. Operators require next-generation PON systems to have a higher capacity, longer reach, larger bandwidth, and the capability of accommodating more users. These operators recognize the economic benefits of investment, rationalization, and enablement of prospective revenue creation opportunities and this scenario serves to create new opportunities for the operators.

Regarding Investment Rationalization

The deployment of next-generation PON will halve the installation cost, while operations and maintenance costs will be reduced by 30%. This results from the intelligent operation and management of the passive nodes. The total cost of ownership is reduced to 50% as compared to pure fiber deployment. Having next-generation PON means fewer backhaul links as individual fibers are required to maintain equivalent capacity.

Regarding Revenue Creation Opportunities

The plug-and-play remote nodes allow the mobile network to provide a potential increase in Internet of Things (IoT) connectivity effectively. Thus, it allows operators to easily grow in the ever-expanding IoT connectivity market.

3.3.3.3 Next Generation PON

Some service providers are already implementing 5G passive optical networks that enable them to expand their business portfolio. For example:

- SK Telecom has signed an agreement with telecom equipment companies including Cisco Systems and Nokia for the global expansion of 5G-PON.
- Verizon, in addition to operators such as Altice (an innovative New York-headquartered international company specializing in the implementation of digital technology for the provision of consumer services) and Portugal Telecom, selected NG-PON2 (Next Generation PON2), as it uses tunable optics. Its multiwavelength approach allows operators to allocate different subscriber types to different wavelengths. It is also a path forward for Verizon because it unites its residential, business, and mobile subnetworks onto a universal network.
- Altice is upgrading its legacy fiber access networks in order to use the full capabilities of NG-PON2 and so meet the needs of the most demanding urban deployments of 5G small cells alongside enterprise and residential broadband demands.
- Huawei has launched Flex-PON 2.0, an edge tool for the internet-driven FTTx solution, which enables the same service board to be compatible with 6 PON technologies, fully reusing OLT resources on live networks. This technology helps the operator greatly improve its network bearer capability and device utilization efficiency.
- Nokia has signed an agreement with China Mobile, China Telecom, and China Unicom to deliver technologies and expertise across China for radio access, core, passive optical networks, IP routing and optical transport network (OTN), SDN, network

management, and professional services worth a combined value exceeding $2 billion.

Fronthaul and Backhaul Networks

Applying a PON into the mobile fronthaul (MFH) has good prospects along with the features of low cost and wide coverage. It has been demonstrated that fronthaul has the largest share of the 5G PON market.

NG-PON is a flexible, reliable, and efficient fronthaul transport solution, which provides a single distribution network for both cellular and high-speed fixed broadband. It is uniquely suited to 5G fronthaul applications and is increasingly attracting widespread industry attention as the 2020s progress toward and through the 2050s.

Comparing existing mobile fronthaul technologies, such as dark fiber, passive WDM, and active DWDM, NG-PON has various advantages such as wide bandwidth, low latency, fiber savings, plug-and-play ONUs and relatively low costs. These benefits make it relatively straightforward for NG-PON to satisfy 5G fronthaul requirements such as dense site deployment, growing bandwidth needs, and shorter latency. It also reduces the size of the remote unit at the last mile and supports flexible installation/configuration of the fronthaul network.

The implementation of PON network topology for dense 5G fronthaul networks can substantially save optic fiber deployment, can share the existing space of fiber infrastructure setting and room space, and can also save significantly on the costs of network construction and maintenance. This is why several players implement PON for 5G fronthaul network. For instance:

- In 2018, ZTE was the first vendor to complete the development and testing of 10G WDM-PON equipment for the fronthaul device interface requirements of China Telecom's 5G trial in Shanghai.
- SKT developed an innovative fronthaul (transport) network solution called 5G-PON that reduces the size of the remote unit at the last mile and supports flexible installation/configuration of the fronthaul network. This solution saves electrical energy and reduces transport network costs for the mobile operator.
- Nokia Bell Labs have announced the first successful demonstration of ultra-low latency 10G PON for mobile fronthaul.

PON is an excellent fit for mobile backhaul traffic, particularly for macro cells and small cells located in urban areas. PON used for the mobile backhaul network has substantial benefits including very wide bandwidth and low operating expenses. However, clock synchronization is a major issue, which has to be addressed, because base stations (BSs) within the network have to be synchronized in order to provide high quality of service (QoS) to customers. It is clearly good timing to deploy NG-PON to backhaul 5G networks as capacity requirements on some 5G networks will be significantly higher than the current (legacy) 4G networks. NG-PON can support four distinct full-duplex 10-Gbps feeds and each wavelength uses Ethernet to carry 5G packet information, which is necessary for 5G backhaul.

NG-PON is progressively gaining momentum in 5G backhaul, ensuring a future-ready investment, opening-up a new direction for operators. Examples include:

- Adtran upgrades NG-PON2 for 5G as optical backhaul gains support.
- Verizon has selected NG-PON2 for its 5G backhaul needs.

The ITU-T and IEEE organizations are creating next-generation PON standards to support the increasing bandwidth requirements of emerging services. Important approaches can be segmented into two phases, namely:

- *NG-PON1:* An intermediate upgrade for Gigabit-PON (GPON) systems, supporting 10 Gbps on the downlink and 1 Gbps on the uplink. The NG-PON1 system is basically an improved TDM PON (compared with GPON).
- *NG-PON2:* 5G will be the most significant driver for NG-PON2. This is a long-term solution using entirely new types of optical network such as WDM-PON. It supports 10 Gbps on both uplink and downlink. NG-PON2 technology will definitely outperform NG-PON1 technologies in terms of compatibility, bandwidth, capacity, and cost-efficiency. Its architecture has been developed on a time and wavelength division multiplexing (TWDM) approach, which stacks four wavelengths in a coordinated manner onto a single fiber with each wavelength delivering 10 Gbps.

Thus, NG-PON2 provides the best technical fit for both 5G backhaul and fronthaul.

The NG-PON2 5G fronthaul network architecture exhibits the following technical features in that it:

- Reduces the amount of fiber needed by 5G fronthaul networks with high site densities.
- Offers a dedicated wavelength for each user with high transmission efficiency and extensive bandwidth resources.
- Uses a colorless ONU technology with the ensuing low cost and easy maintenance.
- Supports a unified optical access platform and online transmission sharing, instantaneously meeting the service requirements of mobile users and home users.
- Uses Auxiliary Management and Control Channel (AMCC) technology, which consists of two mechanisms: RF pilot tone and baseband-over-modulation. Other key technologies include optical modules and protection switching. All of these technologies are now playing an important role for deploying WDM-PON in 5G, and their protocols will be formulated by the related standards bodies.

3.3.3.4 Gigabit Ethernet Passive Optical Network and CATV-Related Standards

Standards known as Gigabit Ethernet PON (GPON) and Ethernet PON (EPON) are increasingly important.

GPON provides three layer-2 networks: ATM for voice, Ethernet for data, and proprietary encapsulation for voice. It also provides 1.25 Gbps or 2.5 Gbps downstream and upstream bandwidths scalable from 155 Mbps to 2.5 Gbps. One example is Ericsson's BLM 1500 GPON.

EPON uses Ethernet packets instead of ATM cells. It employs a single layer-2 network that uses Internet Protocol (IP) to carry data, voice, and video. It usually offers 1-Gbps symmetrical bandwidth, making it particularly popular in modern networks.

A wide range of DWDM standards have also been introduced by the ITU, related to the application of DWDM for GPON. These standards are expanded upon in detail in Chapter 6 (CATV) but as they relate to GPON they are briefly introduced here:

"GPON" (or EPON);
10G PON (10G EPON)*;
25G PON;
50G PON.

In all instances, G means 1 Gbps. GPON coexists with a standard termed XGS-PON.

Each PON standard is assigned to a specific wavelength according to ITU PON wavelength plans.

3.3.4 Advanced 5G-Related Pluggable Fiber-Optic Transceivers

The twenty-first century has entered the digitally advanced phase in which 5G technology is already abundantly available and with technology appropriate to the requirement for higher base station density, there is ever-increasing demand and market space for high bit-rate optical transceivers. Through the earlier part of the twenty-first-century LTE base stations of the 4G level mainly used 10G optical transceivers, but 100G optical transceivers had already gained acceptance in the front-end 5G optical transmission. With the construction of large-scale data centers and greatly increasing traffic density, the demand for optical transceivers in the data center had already been shifted from 10G/25G to 40G/100G, and since 2017 100G has become the mainstream de facto standard.

- *10G:* These optical transceivers communicate using only one fiber with two different wavelengths. On one side of the transceiver, the wavelength is transmitting the 1,330-nm signal and receiving the 1,270-nm signal, whereas the other side has the mirror image, so the optical transceiver will transmit a 1,270-nm signal and receive a 1,310-nm signal. It enables the use of the Ethernet for transporting data, voice, and video traffic within an enterprise and a carrier's network. The key advantage with this system is that it can double the speed compared to a traditional optical fiber pair. By using a 10G uplink, it can be arranged to ensure nonblocking communication between the edge switch and the core network. Therefore, 10G ensures the presence of a nonblocking bandwidth in the network.

- *100G:* 100G is well-established in long-haul networks, and the industry focus has shifted to optimization and deployment of 100G in metro networks. The wide demand for 100G optical transceivers is driving advancements in optics, high-speed lasers, and 100G integrated circuits (ICs), which can cause the price of 100G equipment to decline over time as volume manufacturing increases, especially additive manufacturing.

Various metro-optimized WDM platforms were introduced in the market in 2016 and 2017. Many of these platforms include support for 100G/200G CFP2-ACO pluggable coherent transceivers, which has provided a pay-as-you-go pricing model, which is attractive to metro deployments. The next generation of high-density, metro-optimized 100G/200G CFP2-DCO coherent pluggable modules have been shipping in volumes for several years (Acacia, Ciena, Lumentum, Oclara), which adds another element of flexibility to metro-optical deployments.

100G uses 25G laser chip technology which is driven by technological upgrading and cost reduction. Based on various packaging methods these optical transceivers are CFP/CFP2/CFP4, CXP, and QSFP28, which is the current generation of 100G optical transceiver packaging and has now become the mainstream packaging of optical transceivers.

100G optical transceivers have different models and standards, which have become the top-ranking products in large-scale data centers and the telecom 5G market. Thus, content providers, telecommunications carriers, IT hardware/software players, and hosting-based companies all own and operate data centers.

As the number of data centers in peripheral areas grows, the need to interconnect them is typically fueled by the "lateral/metro" networks and other high-speed infrastructure extensions. These factors are also driving upward 100G optical equipment sales. For example, the North American data center market is upgrading 40G to 100G (and higher) and availability of the current high-speed optical transceiver is the main growth driver. Following commercial introduction into the 5G network, the telecommunications market for high-speed optical transceiver demand will be larger than the data center market.

- *200G:* 200G replaces 100G as the relatively new standard for long-haul applications, which is increasing optical transport bandwidth. It allows enterprise customers to achieve performance monetization and improve customer experience. The new technology is gaining in popularity because of its inherent programmability, and the fact that it changes the unit of currency for capacity that is deployed in the existing networks. The 200G multiprotocol multirate solution forms optical transport networks with integrated layer-1 optical encryption configured per client or uplink, as well as providing secure high-capacity transport using a single wavelength. It is modular and enables easy and cost-effective service rollout or expansion of the existing network capacity. It will also facilitate horizontal and vertical network resource slicing for customers, latency-based routing, online fiber monitoring, and T-SDN-enabled networks for centralized controller, resource, and service visualization. A prominent South Korean service provider is using optical transport systems to upgrade its 200-Gbps backbone to facilitate the higher traffic demanded by its 5G wireless networks. OTNs and automatically switched optical networks (ASONs) are characterized by optical transport systems supporting 200 Gbps and enabling a smooth migration to 1 Tbps by the late 2020s. These networks carry traffic using a variety of optical modulation formats, including 8QAM, 16QAM, QPSK, and others to meet the capacity and distance requirements.
- *400G+:* The relatively new 400G is becoming the trend for next-generation optical networks. It is a combination of high-speed connectivity services, increased fiber capacity, and fewer wavelengths to manage, making it an attractive proposition for the network providers, OTT service providers, and web-scale giants. 400G-related optical market driving forces relate mostly to the implementation of networks within the mega data centers and interconnects between these data centers.

400G optical transceiver products are compliant with Ethernet, Fibre Channel, SONET/SDH/OTN, and PON standards. As the designation signifies, these usually operate at data rates of 400 Gbps and higher. They are capable of supporting link distances ranging from very short reach within a data center to campus, access, metro, and long-haul reaches and they are positioned to minimize jitter, electromagnetic interference (EMI), and power dissipation. The emergence of 400 Gbps is reshaping the data center and data center interconnect (DCI) optical landscape. The 400G optical transceiver market is being fueled by the use of mega data centers that implement broadband networks in cloud computing environments.

Videos, internet adoption, and tablets drive demand for broadband mega data centers. Such centers that support online commerce, streaming video, social networking, and cloud services for all industries are expected to implement 400G, 600G, or 800G optical transceivers as their ultimate technology (with 1,200G and 1,600G on the horizon). Software-as-a-service (SaaS) is a major service offering that will leverage 400G optical transceivers in the mega data center. The fast-growing metro WDM market and the need to maximize capacity for short reach DCI applications are likely to drive the 400G (and higher) market. This also aids customers to upgrade the unit of currency for capacity across their network, and Table 3.2 indicates optical transceiver products available from some of the main OEMs.

XR Optics (originally promoted by Infinera several years ago) is emerging as an important standard within the pluggable scenario. Compatible with point-to-multipoint (Pt-MPt) requirements, XR Optics is very important and offers the following advantages:

Table 3.2
Optical Transceiver Products Around 2023

OEM	400G	600G	800G	1,200G
Ciena	Wavelogic 5 Nano		Xtreme 800G	Coming
Cisco Systems	QSFP DD			
Huawei		OptXtrans DC908	600–800G	Coming
Infinera	XR Optics		ICE6 800G	
Nokia	PSE-3c	PSE-3s 600G		

Note: Most products are available in either QSFP or QSFP-DD pluggable format and this applies to all 400G transceivers.

- It is fully compatible with existing infrastructure.
- Its introduction reduces network complexity dramatically, typically by one-half.
- It offers space and power savings.
- Typically over 45% of CAPEX is saved by adopting XR Optics.
- It is an important ongoing development within NG PON.

The Open XR Forum [3] has existed for several years, and benefits from an ever-growing membership including, for example, among carriers: AT&T, BT, Telefónica, and Verizon, and, among equipment suppliers, Infinera and Lumentum, both of whom are collaborating in this field.

3.4 FREE-SPACE OPTICAL LINKS

It is well known that light waves will travel through atmosphere and also through free space. In the 1990s, I actively supported one venture that planned to go commercial in the Earthbound aspect of this type of technology. The details of this technology were really very basic as it almost always relied on LED-based transmitter technology, which greatly limited its maximum transmission speed. For ranges up to a few hundred meters (maximum), it solved a few requirements, but performance was very limited. Another factor was the transceivers were mostly of Russian manufacture. This would have been totally acceptable in the 1990s, but is almost unachievable nowadays.

During the twenty-first century, the development and application of FSO links are certainly alive and growing in importance. In his book [4], Geoff Varrall provides several substantial observations regarding this increasingly important technology, notably Section 7.14.

In this second edition, the subject of FSO is discussed in more detail in Chapters 4, 9, and 10.

References

[1] Nellist, J. G., and E. M. Gilbert, *Understanding Modern Telecommunications and the Information Superhighway,* Norwood, MA: Artech House, 1999.

[2] https://www.ericsson.com/en/blog/2023/1/simpler-rollout-with-tunable-optics (Michael Gronovius).

[3] www.openxrforum.org.

[4] Varrall, G., *5G and Satellite RF and Optical Integration,* Norwood, MA: Artech House, 2023.

[5] Edwards, T., *Gigahertz and Terahertz Technologies for Broadband Communications,* 1st ed., Norwood, MA: Artech House, 2000.

4

DEFENSE SYSTEMS

4.1 DEFENSE—WE ALL STILL NEED IT

4.1.1 The Distant Hope and the Current Reality

When World War II was over, the hope (perhaps even the expectations) of the international community would somehow arrange things so there would be no more war: BBC's moto "Nation shall speak peace unto nation."

Immediately following World War II, the then-Soviet Union (the Union of the Soviet Socialist Republics (USSR)) determined the details of its western border with the rest of Europe and the Cold War began, only to end somewhat abruptly in 1990.

Meanwhile, the world endured the Korean War, the Egypt-Israel conflicts, the Vietnam War, several other wars (often in the Middle East), and, at the time of this writing, Russia's war against Ukraine and the Israel-Hamas war. Unfortunately, deep conflicts between nations (or mad dictators, or terrorism) often lead to war. Over the decades, major nations have responded to this situation by evolving what is often called the military/industrial complex. This generally works as follows:

- At a government level, a serious and violent situation is perceived to have occurred and the appropriate department or ministry informs the head of state. This usually comprises a potential international threat and therefore an international response to this situation is deemed to be necessary.

- On occasion, this might amount to the declaration of a full-scale war requiring Tier 1 corporations to supply equipment called for by all military services (Army, Navy, Air Force, Marines) with the necessary hardware and software.
- Directives are issued and these may include identifying arms suppliers to provide the mainly western "free-world friendly" countries under attack with appropriate weapons (at no cost) and/or financial assistance together with the necessary training.
- Major military equipment includes tanks, other military land vehicles, navy vessels and aircraft (manned as well as drones), SATCOM, and very important land-based weapons such as multilaunch rocket systems (MLRS).
- In the present context, the following are also vital: communications systems, electronic warfare, and radar systems, plus the means to electronically attack enemy systems.

Some details are now provided relating to free-world defense expenditures and communications-related data. The effects of Russia's war against Ukraine, Israel's war against Hamas, and China's stance regarding Taiwan are accounted for leading to the following key data:

- Global (free-world) defense expenditure expected to exceed $3.1 trillion by 2029;
- Global (free-world) defense electronic systems expenditure expected to exceed $86 billion by 2029:
    - Largest segment: military communications (especially space);
    - Fastest (market) growth segment: electronic warfare.
- Global military radar systems expenditure expected to exceed $40 billion by 2029;
- All free-world countries will always have increasing defense budgets.

4.1.2 Defense Data: Focus on North Atlantic Treaty Organization Countries

The aim of this section is to provide relevant data and comments applying mainly to a selection of just 12 NATO countries. North Atlantic Treaty Organization (NATO) member countries total 31 at the time of this writing, and this number is likely to rise to 32, assuming that Sweden joins in 2024. The 12 are selected on the basis of defense expenditure as a percentage of each country's GDP. One specific source of background data was the CIA website [1] because some of the required data points are available on that site. Additional information came from directly searching the internet. The countries included here were chosen because of their percentage of GDP prominence and are Bulgaria, France, Germany, Italy, Latvia, Lithuania, Poland, Spain, Turkey, Israel, the United Kingdom, and the United States. The countries of Bulgaria, Latvia, Lithuania, and Poland were included because of their proximity to the Russian Federation. Turkey is included because it is arguably the most complex example (although just how reliable is Turkey as a NATO member state?). Israel is included, although it is not a NATO country, but it has an exceptionally strong military presence with three Tier-1 players. The data points are provided in Table 4.1.

For many years, the United States has led the NATO pack regarding total annual defense expenditure in absolute terms and also as a percentage of this country's GDP (around 3.6%). It is noted there are several other NATO countries contributing at least 1% of GDPs, but only Bulgaria contributes the 3.6% figure noted earlier.

4.1.3 Some Relevant Defense Abbreviations

However, there is another extremely important element that characterizes defense and military operations. Arguably, this is at least as significant as the military theater itself and here we are talking about the vital surveillance, monitoring, and analysis activities. As any effective manager knows, without accurate and up-to-date information, no task at all can be effectively performed. This is true for both civilian and defense activities, but the main difference is that, in the defense instance, entire cultures can be altered, often irreversibly, and people will die and become maimed if the campaign lacks constantly updated and credible information.

Table 4.1
Defense Data Concerning Selected Countries (NATO Plus Israel, 2020s)

Country	Annual Defense Expenditure (Billions of U.S. Dollars)	GDP (Billions of U.S. Dollars)	Defense Expenditure as a Percentage of the GDP	Recent Annual Change
Bulgaria	3	84.3	3.6	3.2%
France	63	2,800	2.2	Small increase
Germany*	61	4,300	1.4	1.5%
Israel†	23.6	525	4.3	4.2%
Italy	34.8	2,320	1.5	Small increase
Latvia	1.0	5.75	1.7	1.74%
Lithuania	2.07	1,035	0.2	6.0%
Poland	28.1	1,220	2.3	Small increase
Spain	18.5	1,850	1.0	Small increase
Turkey‡	36	2,400	1.5	Small increase
United Kingdom	70	2,800	2.5	0.2%
United States	729	20,250	3.6	Very small increase

Notes:
* The data in this table excludes the extra $100 billion added to this country's defense budget in February 2022 in response to Russia's invasion of Ukraine.
† At 4.3% of GDP, Israel's proportionate defense expenditure is among the world's highest known levels.
‡ Geographically midway between Russia and the Middle East and having a lengthy (Black Sea) border with Russia, Turkey is in a politically awkward position in defense terms.

It seems reasonable to expect every NATO Member country to provide at least 2% of their GDP for their contribution toward defense. Currently, Germany, Italy, Spain, and Turkey come in below that 2%. Although not included in Table 4.1, in July 2022, Finland and Sweden became de facto NATO Member states, and in June 2023 Finland became the thirty-first full Member of NATO.

These requirements lead to the following concepts and abbreviations. There are many abbreviations relating to communications in the defense context, but most are based on C4ISR (Command, Control, Communications, Computers, Intelligence, Surveillance, and Reconnaissance). The first four Cs account for the first four functions: command, control, communications, and computers. This abbreviation dates to the mid-to-late twentieth century. Later, as coding and security requirements gained even more importance, an extra C (for cyber) was introduced, forming five Cs: C5ISR (Command, Control, Communications, Computers, Cyber).

Subsequently, it has been recognized that, being military terminology, an attack-related term should also be added, leading to C6ISR (Command, Control, Communications, Computers, Cyber, Combative). The additional term is combative. Obviously, there are now six Cs in this defense-related abbreviation.

The overall total global free-world markets for all C6ISR equipment exceeds $100 billion and will likely reach around $160 billion by 2030. A similar figure is forecasted to apply to the digital battlefield.

4.1.4 Surveillance, Monitoring, and Networks

During the final decades of the twentieth century, various international agreements, particularly within NATO and what is often known as "UKUSA," ensured that these surveillance, monitoring, and analysis activities expanded steadily. Until 1990, this was also true of the USSR, but with the demise of this previous "second superpower," there inevitably came the downfall of the earlier Warsaw Pact security agreement. From early in the twenty-first century, the world became burdened with a new and highly aggressive Russian leader named Vladimir Putin. Under Putin's regime, on February 22, 2022, Russia summarily invaded Ukraine and, at the time of this writing, this war is still very much ongoing.

Surveillance, monitoring, and analysis are all absolutely essential for the effective pursuit of all modern defense operations. Such activities cannot wait until the outbreaks of active hostilities but instead are perpetually under way; there is constant and persistent electronic international security activity occurring at all times. When things heat up, as, for example, with the Russia-Ukraine situation described above, these ongoing security systems just become busier than ever. Later in this chapter, some details are provided about the types of communications systems involved in such security installations. Historically, such networks derive from what has become known as signals intelligence (SIGINT) or communications intelligence (COMINT).

This then sets the scene for the defense communications systems requirements for today and tomorrow.

It is worth observing that, while cruise missiles travel at only moderate speeds, contemporary fighter jets such as F-16s, F-35s, and Stealth Bombers move at supersonic speeds and often with fractional-seconds time to target. Hypersonic vehicles and many "smart bombs" fly at several times the speed of sound, frequently several thousands of miles per hour. Also, vast and ever-increasing amounts of data must be transferred between nodes, both mobile and static, in ever-shortening time spans. All these aspects point toward an expanding requirement for very high-speed defense communications systems.

In addition to this high-speed systems requirement, defense communications networks must always be robust and this is reflected in terms of the network topologies. Nodes must be accessible via several different alternative paths or routes. Relatively simple axial or pyramidal topologies, while suitable historically, are unacceptable today and, instead, matrix or grid forms must be implemented. In Figure 4.1(a), the upper network represents an axial or pyramidal topology where a failure in early links results in total or at least major breakdowns in the entire structure. In contrast, with the matrix network shown in Figure 4.1(b), signals may become routed over several alternative paths and therefore a failure in one or more paths is overcome (using software control) by the fact that the signal has at least one alternative route. In both cases, COC means chain of command.

In practice, the various nodes may be within terrestrial, tropospheric, or satellite configurations and many modern defense networks are implemented with appropriate combinations of such approaches. Terrestrial and tropospheric communications are covered in Chapter 8, whereas satellite communications is the subject of Chapters 7 and 9.

Actual technologies may therefore be copper or fiber cables or they can be implemented using microwave/millimeter-wave carriers (mmWave). Obviously, with many terrestrial systems, whether static, semistatic, or mobile (land, sea, air, or space), wireless links using microwave or mmWave carriers are often desired and this is also true for satellite systems. Highly secure cabled links almost always employ fiber-optic cables and encrypted digital signals. This applies to defense as well as many commercial satellite ground station links because the alternative of coaxial copper cabling embodying a braided outer conductor is inherently insecure due to radiation and vulnerability to tapping.

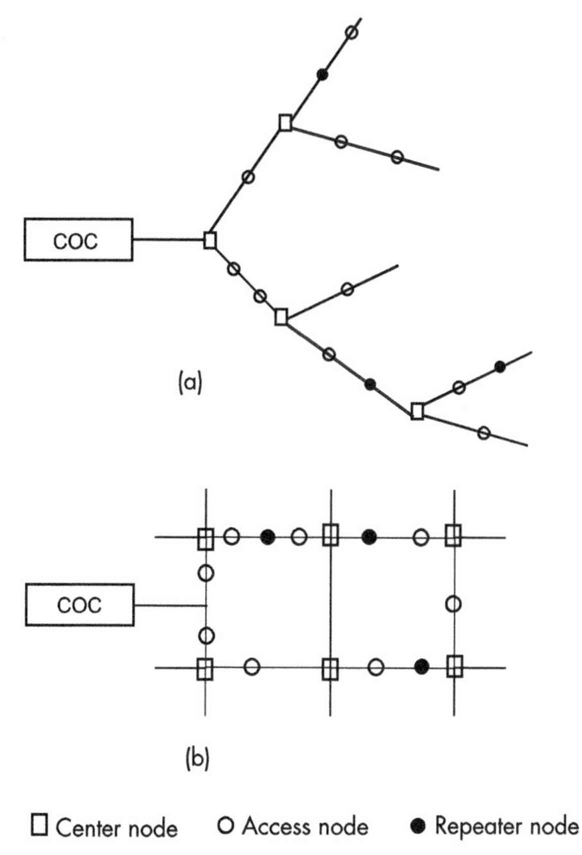

Figure 4.1 (a) Pyramidal or (b) matrix network technologies. *From:* [5].

Some military communications systems incorporate microwave (or mmWave) radio, tropospheric scatter, and a satellite link integrated within one overall network. For example, a typical tropospheric and satellite services radio (TSSR) operates using a 15-GHz carrier on line-of-sight integrating Tr-Tac equipment and ground mobile forces (GMF) satellite terminals.

Digital traffic can be carried, typically up to 6.144 Mbps, using a pseudo-nonreturn-to-zero (NRZ) signal that is also channeled over fiber-optic cables. For the line-of-sight 15-GHz radios, antenna diameters of either 305 mm (31-dBi gain) or 610 mm are employed and installation time is generally less than 1 minute. Flat-panel antennas are increasingly penetrating the market for these types of systems.

The system operator chooses either horizontal or vertical polarization for optimum performance over these types of links.

4.2 MILITARY SATCOM

4.2.1 Some Background

Ever since the shock launch of Sputnik by the USSR in 1957, it became clear that satellites were going to become a very important element on the global scene. After this seminal event, satellites such as Early Bird and Telstar became well known and successful within the scope of technology in that bygone era. The first satellite TV (1960s vintage) delivered grainy received images that lasted a matter of minutes because the spacecraft were in LEO. Interestingly, in this digital age:

- LEOs are once again of great importance (see Chapter 9 in particular).
- Trios of satellites, each in GEO orbit are also very important for TV transmission and some military applications.

An ever-expanding variety of defense satellite systems and associated spacecraft are being deployed on a global basis. The United States and Western Europe remain the principal sources of space hardware, and nations within Europe, often with companies operating in consortium-like groups, are in the forefront of this technology.

In defense terms, Skynet refers to a sequence of military satellite systems implemented by the U.K. Ministry of Defence since the 1970s, starting with Skynet 1. British Aerospace (BAE Space Systems) was the prime contractor for Skynet 4, and the Franco-British joint venture then known as Matra Marconi Space was responsible for the design and build of the communications module. A schematic illustration of a group of Skynet ground stations, based upon a photograph of the RAF (U.K.) Colerne installation, is provided in Figure 4.2. The large, high-gain, dish reflector antennas provide secure communications links for U.K. armed forces and the NATO alliance. The final three Skynet 4 spacecraft were launched over the 1999–2000 period and a new system is currently being planned.

Skynet ground stations comprise three-dish specialized VSATs and a schematic example is shown in Figure 4.2.

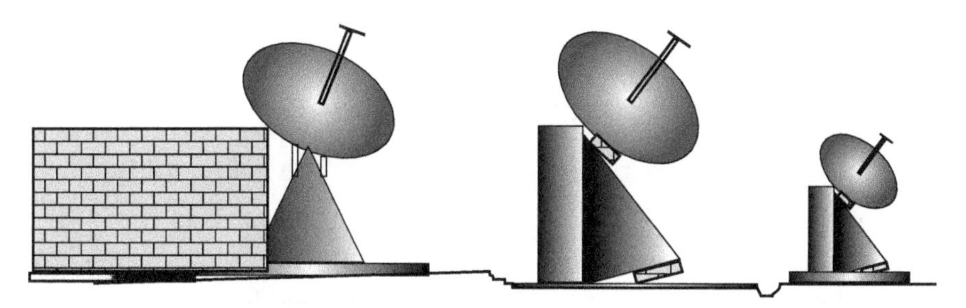

Figure 4.2 Skynet ground station schematic. *From:* [5].

Following Skynet 4, the next-generation U.K.-led system (Skynet 5) involved a radically different contractual approach, which was proposed in 1999. The proposition was that Skynet 5 should be developed under a private-public partnership arrangement in contrast with the traditional procurement program arrangements and a 2-year study was set up in the spring of 1999 with the aim of awarding study contracts to two competing teams. The first team, which might be termed the "British-British/U.S.-U.S. Team," comprised BAE Systems (United Kingdom), British Telecom (BT), TRW, and Lockheed Martin—two British firms and two U.S. corporations with a balanced expertise and experience in appropriate fields. The second team, named Paradigm Secure Communications, comprised just two companies: Matra Marconi Space and TRW. Matra Marconi no longer operates.

This was the first time that any government had proposed privatizing a classified satellite program, and doubtless the maintenance of secrecy was the principal challenge facing both government and contractors during all phases including the study and the final production contract. Several factors influenced this approach: cost savings to government and therefore the taxpayer; opportunities for industry during the study, development, and production phases; and the possibility of trading any excess satellite link capacity to commercial customers and/or NATO. The total cost amounted to approximately $5 billion.

Over the decades ranging through the 1990s to the present era, several types of military space systems, frequently called MILSATCOM, were brought into operation. Specialized NATO systems were also implemented.

4.2.2 Further Military Space Developments

4.2.2.1 Current and Near-Future Developments

SATCOM requires the application of a rigorous process of certification that typically takes 18 months to 2 years. The main requirements for the spacecraft are [2]:

- Although commercial off-the-shelf (COTS) components can be used, these must be thoroughly screened for space environment endurance.
- The selected constellation must have sufficient adaptability and flexibility for the mission.
- There will increasingly be a noticeable trend away from the relatively large parabolic reflector antennas and toward flat-panel arrays.
- The capability for switching between constellations will be an increasing element.
- FSO communications are increasingly common.

The SATCOM-based U.S. Army and Navy multiuser objective system (MUOS), which became operational in 2019, is briefly mentioned here. This is a relatively narrowband system and our considerations are with broadband, which is why no further details are provided regarding MUOS.

Very small aperture terminals (VSATs) are often used for military applications because the Earth stations can be readily transportable by land and sea. In general terms, VSATs tend mainly to operate either at Ku or Ka-band. At the time of this writing, China is building a massive military heliport located on the country's west coast, immediately opposite Taiwan. This will almost certainly require one or more high-security VSATs dedicated to military communications for this heliport.

4.2.2.2 The United Kingdom's Skynet 6 Program

Driven by several international wars and threats, including Russia-Ukraine and China-Taiwan, the United Kingdom needs to maintain a military SATCOM capability that is fit for the mid-twenty-first century. The Skynet 5 PFI program ended in 2022, and, 4 years earlier, the

United Kingdom's MoD began defining the specification for Skynet 6A.

Because the Skynet 5 PFI approach turned out being a poor value for money, the Skynet 6A project arrangements centered on the large Europe-wide aerospace company known as Airbus, in particular, Airbus' facilities located in Stevenage and Portsmouth (both in England).

Under an MoD contract worth approximately $600 million, signed in 2020, Airbus is manufacturing Skynet 6A with a planned launch year of 2025 (using a Falcon-9 rocket operated by SpaceX). Skynet 6A will have the following characteristics (improvements compared with Skynet 5):

- Electric orbit raising and station-keeping propulsion;
- More RF communications channels;
- Greater capacity and versatility.

The foundation for Skynet 6A's technology is Airbus' Eurostar Neo satellite bus, which is a flexible system architecture that combines U.K. government, the governments of several allies, and commercial satellites. The manufacture of this spacecraft started in 2021 and an artist's impression of Skynet 6A is shown in Figure 4.3.

Figure 4.3 An artist's impression of Skynet 6A in orbit. (© Airbus. Special thanks to Jeremy Close who facilitated permission to reproduce this image.)

The overall structure of Skynet 6A is what is known as a CubeSAT. This important technology is described in Section 4.5 and in more detail in Chapter 1.

It has already been decided that Skynet 6 must be fully compatible with at least the following U.S. military assets:

- Advanced Extremely High Frequency (AEHF) (a 2021-launched SATCOM system);
- Wideband Global SATCOM (WGS);
- MUOS;
- Unmanned aerial vehicles (UAVs);
- The F35B Lightning II fighter aircraft.

Skynet is already a substantial part of the MoD's "Future Beyond Line of Sight (FBLOS) Satellite Communications Programme," which extends to 2041 and has the following four elements:

- Secure Telemetry, Tracking, and Command (STT&C), to provide for assured U.K. control and management of satellites and their payloads into the future;
- Service Delivery Wrap, a support contract to manage and control the Skynet constellation and ground infrastructure;
- Skynet 6 Enduring Capability, the facility to provide and operate communication satellites and ground infrastructure into the future;
- Skynet 6A (specifically known as a Single Transition Satellite).

On July 3, 2020, the U.K. government stated that it had spent $600 million in U.S. dollars to acquire a 45% stake in the OneWeb LEO SATCOM company (see Chapter 8). This investment includes what is known as a "golden share," which provides for control over any future ownership that may take place. It is considered likely OneWeb will be incorporated into the Skynet 6 architecture. OneWeb satellites are already manufactured by a joint venture including Airbus Defence and Space, which places the current Skynet operator in a strong position regarding its future involvement in Skynet 6.

In February 2023, the U.K. company known as Babcock International won the $500 million Service Delivery Wrap support contract to operate and manage Skynet 6A ground infrastructure and new user terminal integration. The contract began operating in spring 2024 and will last for 6 years. This new Service Delivery Wrap contract will have to remain in place at least until 2028, when yet another new generation of military SATCOM is expected to begin. It is anticipated the total cost of this transition will be in the region of around $7 billion.

4.2.2.3 The U.S. Space Force

The earliest concept of a U.S. Space Force (USSF) was in 1958, just 1 year following the launch of Sputnik. In that year, President Dwight Eisenhower led the initial discussions on this important concept. However, the first military space programs started at the end of World War II (i.e., in 1945). Nine years later and with the Cold War in full swing, the U.S. Air Force gained a new division known as the Western Development Division (WDD), which was the world's first dedicated space organization. In 1982, the WDD became efficiently unified to form Air Force Space Command, and this unit performed vital support for American (and allies) conflicts including the later stages of the Vietnam War and through the Gulf War.

Prior to these developments, the USSF almost began in 1982 when Ronald Reagan's administration initiated the Strategic Defense Initiative. Seventeen years later, the U.S. DoD considered that a Space Corps should be started, but the events of September 11, 2001, and the subsequent War on Terror took precedence, and it was not until December 20, 2019 that the U.S. Space Force Act became signed into law, hence establishing the U.S. Space Force as the first new independent military service for many decades.

Today the USSF is the world's only dedicated independent space force, and, with around 8,600 military personnel, it is also the smallest division of the U.S. Air Force, summarizing the order of responsibilities:

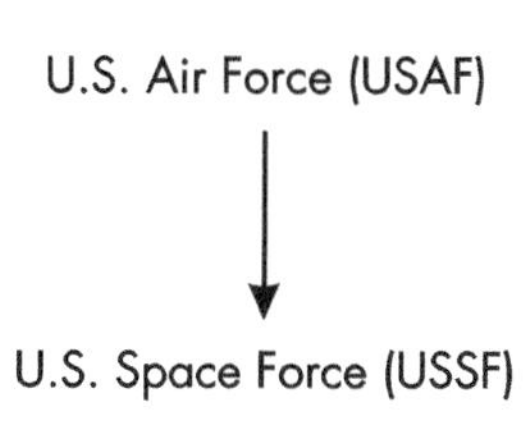

The USSF has the following features:

1. It is the space service branch of the U.S. Armed Forces.
2. It is the world's only dedicated independent space force.
3. Together with the USAF, it is a part of the Department of the Air Force, which is led by the Secretary of the Air Force.
4. The military heads of the USSF are the Chief of Space Operations and the Vice Chief of Space Operations.
5. It operates a total 77 spacecraft, covering:
   (a) GPS;
   (b) "Space Fence";
   (c) Various military communications satellite constellations;
   (d) X-37B "spaceplanes";
   (e) The U.S. missile warning system;
   (f) The U.S. space surveillance network;
   (g) The Satellite Control Network.

The USSF is also responsible for the equipping, organizing, and training space forces toward certifying them fit for operational employment.

4.2.3 Relevant Optical Systems

During the early decades of the twenty-first century, various organizations became increasingly interested in optical communications. This trend very much includes defense requirements that require a deep and strong expertise in optical line-of-sight systems. Typical line-of-sight optical links are illustrated schematically in Figure 4.4.

Laser transmitters and photodetecting receivers are located in the ground station and also on the satellites. The light beams are modulated at the transmitters and demodulated at the receivers. One intersatellite link (ISL) and one satellite-to-ground link are shown in Figure 4.4, and some further technological details were provided in Chapter 2. Relatively wide operating bandwidths are possible using optical links compared with radio. Optical concerns terahertz frequencies (terabits per second transmission), whereas radio is mostly focused on gigahertz frequencies (gigabits per second transmission).

Over the very long ranges (typically thousands of kilometers), an important issue known as scintillation or atmospheric turbulence

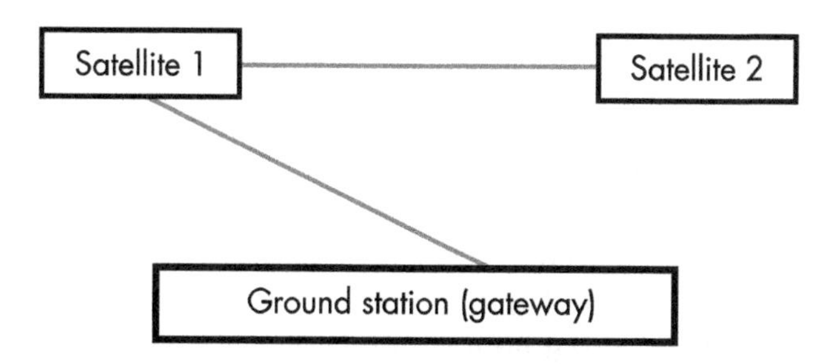

Figure 4.4 Schematic of two optical line-of-sight (LoS) links (the thin orange sodium lines are the beams).

requires the application of adaptive optics to (largely) counteract this effect.

A facility known as the Starfire Optical Range (SOR) [3] represents a good example of a leading-edge research organization in terms of adaptive optics research. SOR is also a secure military site, owned and operated by the USAF and located at the Kirtland Air Force Base in Albuquerque, New Mexico, United States. According to its website, SOR's primary duty is to "develop and demonstrate optical wavefront control technologies." It is interesting to note the use of the term "wavefront" because this indicates a very fundamental description of this highly significant challenge. The purpose of Starfire is to conduct research to use adaptive optics to remove the effects of scintillation or atmospheric turbulence, which otherwise seriously interferes with the performance of any free-space optical link. When not compensated, turbulence interferes with laser beam integrity over distances. Lasers are being used for long-distance, high-bandwidth communications, and accuracy in air-to-air laser connectivity is important for data integrity. Scintillation is also a problem in the development of weaponized lasers, such as the airborne lasers under development for the interception of intercontinental ballistic missiles.

Adaptive optics (AO) is a technique comprising the precise deformation of a mirror in order to compensate for light distortion. It is used in astronomical telescopes and also in laser communication systems to remove the effects of scintillation (see above), in optical component fabrication and in retinal imaging systems. The exact distortions are measured and stored in a computer. The AO technique comprises the precise deformation of a mirror or an LCD array (according exactly to the stored distortion data) so as to compensate for the light distortion.

SOR's optical equipment includes a 3.5-m telescope, which is "one of the largest telescopes in the world equipped with adaptive optics designed for satellite tracking" according to the USAF, a 1.5-m telescope, and a 1-m beam director.

The SOR facility has also demonstrated the transmission and reception of a group of three green light beams converging on a spot point on a satellite. This is shown schematically in Figure 4.5.

It is also apparent that SOR is engaged in research into how ground-based laser transmitters may be used to disable satellites, effectively an antisatellite weapon.

These systems also enable high-accuracy ranging and tracking.

4.3 GLOBAL SECURITY SYSTEMS

The National Security Agency (NSA) was gradually developed by the United States during the years of the Cold War, the 42-year standoff between what were perceived by most as being the world's two superpowers of those times: the United States of America and the USSR. With electronic intelligence gathering (ELINT) being widely recognized as critical in modern warfare—both cold and hot—it was completely understandable to find the protagonists evolving sophisticated global eavesdropping facilities.

By the late 1980s, the NSA had grown to become America's most secret, largest, and almost certainly most costly security organization. The NSA has access to land-based, ship-borne, and air-based facilities

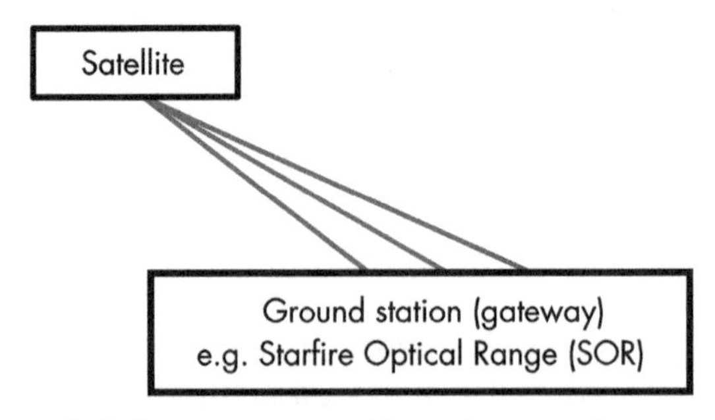

Figure 4.5 Green light beams transmitted from the ground station and focused to a spot on the satellite.

and uses satellites together with other communications systems to channel information back to its intelligence bases. Associated mobile security systems are located on board ships, submarines, aircraft, and spacecraft globally.

The NSA has built up intelligence links with the following four close allies: Australia, Canada, New Zealand, and the United Kingdom. The resulting secret agreement is called UKUSA, and all five flags of these nations are frequently displayed at each UKUSA location—and definitely on days of special VIP visits. UKUSA refers to the high-level official agreement between these close allies to exchange intelligence information on what amounts to a continuous basis. The NSA at Fort Meade in the United States is Anne Arundel County's largest employer and it is also one of the largest employers in the entire state of Maryland. This organization implements information encryption techniques that are second to none internationally, and the term *cryptologic*, referring to cryptographic systems, is frequently used across and within the NSA.

More standard terms include SIGINT, meaning electronic signals intelligence, and also INFOSEC, which is shorthand for information security. Most buildings located in the NSA facilities are windowless for additional security. Walls are well shielded against stray electromagnetic radiation, using internal metal coatings, and the windows themselves are unable to let radio or optical signals either in or out, so a high level of security is essentially guaranteed.

All forms of electronics, including very high-speed digital communications systems, are of vital significance to the NSA and also essentially all other security organizations. In order to receive sensitive low-power signals from many sources, quite large satellite signal receiving dishes are essential, covered with white radomes. The radomes protect from the worst effects of bad weather, which can, at worst, cause signal loss, help keep the important parts of the dish clean, and also stop any onlooker from seeing the direction in which the antenna is pointing—important where security is paramount. The radomes are usually made of a neutral white outer surface color. White reflects the Sun's rays very efficiently and limits the extent to which the system heats up. On the whole, the hotter the satellite receiver, the worse the reception quality because background noise increases. In all probability, the security people would prefer their radomes to be camouflaged so as to merge with the natural surroundings, but, in the end, a white surface wins out.

By 2000, the Menwith Hill NSA base enjoyed the benefits of some 29 radomes in total, and it is considered likely that several more have been added during the twenty-first century to date. Soon optical links (see Section 4.2.3) will very likely be included, although, for obvious security reasons, wavelengths will likely be outside the visible ranges. Quantum key distribution is then required for security purposes.

An elevational view of the main sections of the base is provided by Figure 4.6, which is founded upon an actual photograph published in the press during the late 1990s.

Several differently sized radomes are clearly visible in Figure 4.6 as well as windowless buildings and other structures. The base is surrounded by the open countryside of North Yorkshire County, England. According to well-publicized media reports, entry into the most recent facility at Menwith is controlled by a retinal imaging technology in which only those staff for whom matching (stored) retinal images precisely correlate gain access.

An aerial view, again based upon an actual late 1990s photograph, is given in Figure 4.7.

There are some additional radomes that are not visible in this particular view. Areas comprising collections of many small buildings are indicated to the left background—these are mainly administrative together with some residential accommodation. Further rectangular blocks house the operations elements of the base and these are much closer to the radomes. A relatively insecure dish reflector antenna, without the benefit of a radome, is visible in the foreground.

The radomes, in any operational base, are by no means randomly located. Some groups of radomes are especially interesting because

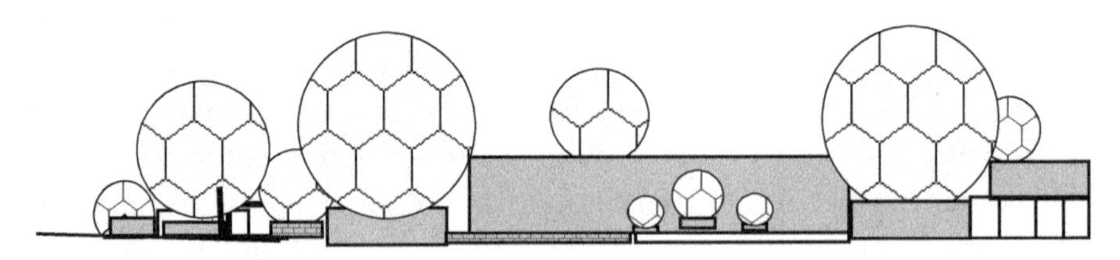

Figure 4.6 A westerly elevational view of the Menwith Hill NSA Base. *From:* [5] (based upon an actual photograph).

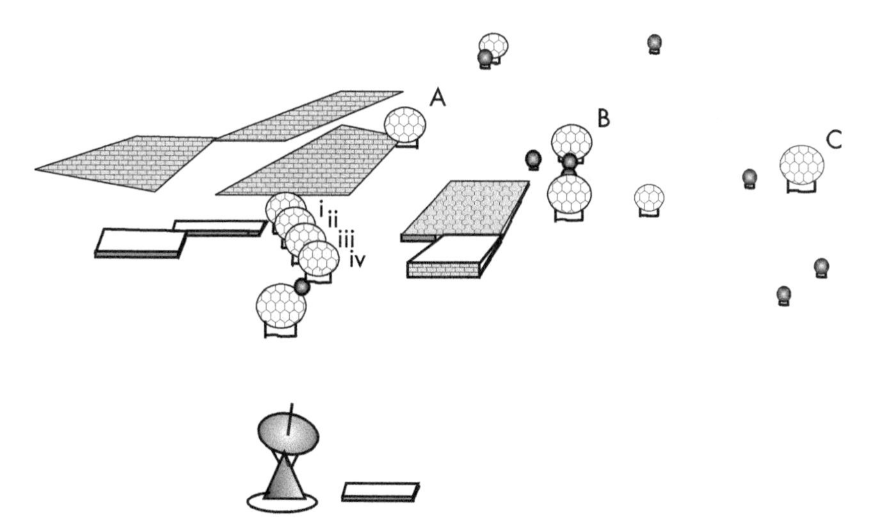

Figure 4.7 An aerial view of the Menwith Hill Base antennas and buildings. *From:* [5] (based upon an actual photograph).

they appear to comprise large phased-array configurations. In Figure 4.7, two such groups are identified: groups A, B, and C, and also groups i, ii, iii, and iv. The radomes within each of these groups fall into distinct lines—that is, they form linear rows and it is well known that these types of configurations usually comprise phased arrays. The best-known nondefense example in the world is probably the Very Large Array (VLA) in Socorro, New Mexico, United States, that is operated by the National Radio Astronomy Observatory and which comprises a total of 27 individual antennas mounted on railroad tracks.

With phased arrays, the individual antennas, within the radomes, are electronically interconnected and the signals are fed to and from each antenna with differing respective phases—hence the term *phased array*. Using this principle, the resultant composite microwave beam, associated with all the elements of the array, is not only combined—it can also be extremely rapidly electronically steered so as to point in different directions. This would have considerable advantages for satellite-based surveillance, including the situation whereby it is impossible to know pointing directions without any knowledge of the electronic steering programming.

In phased arrays, the separation of the antenna elements determines the resolution of the final scanned image—the larger the separation, the greater this resolution—so array A, B, C should have a better resolution than i, ii, iii, iv. It is also interesting to note that the two distinct arrays are disposed approximately 45° from each other. This probably means they track quite different satellite constellations.

Menwith Hill is also a key element in the U.S. Space Force program (see Section 4.2.2.3).

4.4 TYPICAL TECHNOLOGIES

4.4.1 General Aspects

Most of the basic types of products associated with defense installations today are essentially commercially available. It is the final systems configuration—the ultimate systems specification—that determines the defense nature of the installation. It is important to first consider power amplifiers because these subsystems tend to dominate limitations and opportunities. Alternate abbreviations are RFPA (RF power amplifier, which is quite general) and solid-state power amplifier (SSPA), which is specific to solid state or semiconductor. Further details of these important aspects were provided in Chapter 1.

4.4.2 Space Segment Challenges and Advances

The space segment presents systems designers with particular challenges, but this should not be surprising in defense application terms because components and modules fitted for space must generally be space-qualified. In many instances, such components must be "rad-hard," that is, they must remain operational in environments where severe ionizing radiation flux is present, up to a specified limit.

In most current and projected applications solid-state components and modules are implemented in the bulk of the systems. Where relatively high power is required, particularly at higher microwave and millimeter-wave frequencies, tubes such as traveling wave tubes (TWTs) remain essential components. The power capabilities of solid-state components and modules are subject to continuous improvement resulting from new device technologies and advances in power-combining techniques. Monolithic microwave integrated circuits (MMICs) represent a very significant aspect of this scenario. In particular, the Defense Advanced Research Projects Agency (DARPA), under the U.S. Department of Defense, frequently reports on the status of such elements. The situation that applied to power MMICs around the 2023/2024 period was outlined in Chapter 1, and to achieve output power levels exceeding 100W, SSPAs are used as drivers for tubes such as TWTAs. In this case, the signal is first amplified by the SSPA and the output from this MMIC amplifier is then applied to the TWTA input. Typically, an SSPA output power level of 2-W at 26 GHz is followed by a TWTA delivering up to 500-W of output power at the same frequency [4]. These technologies were covered in Chapter 1.

References

[1] www.cia.gov.

[2] "Resilient Integrated Solutions for Government and Military SATCOM," webinar, June 12, 2023, Multi-Layered Security and Resistance, https://www.idirect.net/event/webinar-resilient-integrated-solutions-for-government-and-military-satcom/.

[3] https://afresearchlab.com.

[4] https://elvespeed.com.

[5] Edwards, T., *Gigahertz and Terahertz Technologies for Broadband Communications,* 1st ed., Norwood, MA: Artech House, 2000.

5

DIGITAL TELEVISION

5.1 CONVERGENCE REVISITED: ANALOG VERSUS DIGITAL

Today almost everyone agrees that digitization is the way to go as far as signal transmission and processing are concerned, practically anywhere and certainly including television.

While all real-life signals are initially strictly analog, the digital encoding of these signals at the earliest possible point well ahead of transmission provides immense benefits. This concept makes sense for several compelling reasons:

- Minimization of noise effects.
- Enabling of substantial bit-compression of video and audio signals.
- Enabling error correction to be deployed (which is not feasible with analog).
- Digitization means that identical processing techniques can be used for all signals, whether audio, control, data, or video.
- Encryption is easily arranged.
- Channels can be easily allocated.
- The transmission power needed is generally lower than that for analog.

- Robust modulation schemes can be implemented that combat ghosting and phase error issues.
- True high-definition displays can be implemented (HDTV).
- There is excellent compatibility with most of the burgeoning digital technology that is widely available (e.g., VLSI technology); see Chapter 1.
- There is general compatibility with many of the increasingly digital transmission systems that are being implemented in almost all G8 countries for at least cable and satellite.

5.2 DIGITAL VIDEO BROADCASTING

The generic term covering digital television is digital video broadcasting (DVB). This was initially pioneered with satellite (direct-to-home (DTH)), but is also available using terrestrial broadcast TV, mainly as a result of the advances in digital compression technology.

Figure 5.1 shows three distinct classes of DVB.

But, there are some drawbacks to implementing digital TV; the main ones are:

- The quality of service degrades rapidly as the receiver departs from the edge of the specific service area.
- Terrestrial broadcasting stations need to develop a new transmission infrastructure, although with satellite (DVB-S) no changes to the infrastructure are required unless LEO satellites are used, in which case Ku or Ka-band inputs will enter the antenna.

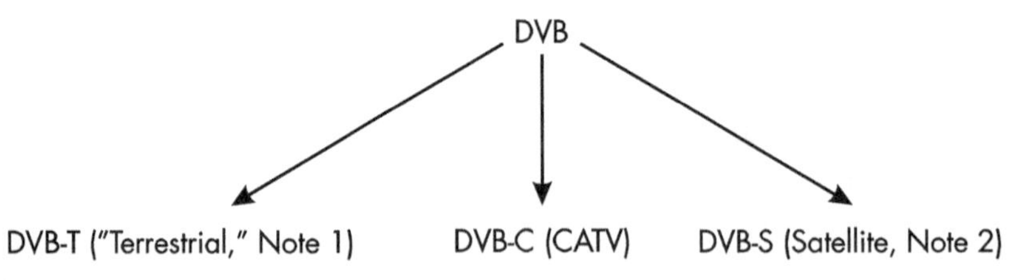

Figure 5.1 The three classes of DVB. Note 1: DVB-T (terrestrial usually via a Yagi antenna but includes internet via fiber). Note 2: DVB-S (satellite most often via a dish-reflector antenna). DVBs C and S are both described in Chapter 6.

- If the received signal-to-noise ratio is too low, a periodically broken picture will result. This is due to increasingly missed bytes of information, eventually causing a kind of dynamic disintegration of the scanned-in frames.

There are additional drawbacks applying to terrestrial TV in general (including DVB-T), in particular, because of the relatively narrow bandwidth fewer channels available and the signal tending to drop out completely under certain specific (anticyclonic) weather conditions.

5.3 MOVING PICTURE EXPERT GROUP

Digital television is an important application of digital signal processing (DSP) and, as such, it requires special formats or standards. The Moving Picture Expert Group (MPEG) represents the most important standards group that has developed and published software known as MPEG2 and MPEG4, both of which involve digital compression and are vital in DVB applications. MPEG publishes standards for the generic coding of moving pictures and associated audio information. At the time of this writing, the latest public release was in 2015, with an amendment dated 2016 (reference H.222.0).

It can be seen from Table 5.1 that there are checks and balances when choosing between the two standards and this can be a very hard

Table 5.1
MPEG2 and MPEG4 Compared

MPEG	**MPEG2**	**MPEG4**
Algorithm	Believed to be 8 × 8 discrete cosine transform (DCT)	16 × 16 DCT, higher rate of compression
Encoding	—	Best in this respect regarding encoding onto portable media such as USB memory sticks
File size and implications	Larger than for MPEG4	Smaller than for MPEG2, works well across mobile networks
Flexibility	—	More flexible than MPEG2
Format and implications	Easier to use than MPEG4	—
HDTV quality	Best; better than MPEG4	—
Interactive features	—	More than MPEG2
Video compression	Easier to use than MPEG4	—

Note that there is no MPEG1 or MPEG3.

exercise. Overall, it looks like MPEG4 may usually be the best choice while sacrificing some HDTV quality. This will be a matter of user experience and acceptably.

5.4 MULTIPLEXING AND CODED ORTHOGONAL FREQUENCY DIVISION MULTIPLEXING

Experience with using the European standard known as coded orthogonal frequency division multiplexing (COFDM) has shown this to be particularly resistant to ghosts and multipath fading. Japan's NHK Science and Technical Research Laboratories also support COFDM.

FDM was commonly used in communications of many forms before the digital age led to a much greater emphasis on TDM. The principle of FDM is basically very simple: the radio spectrum is carved up into a series of adjacent channels or spans of frequencies and different programs are transmitted within specific channels. There are three problems with this basic technique, notably:

1. It is necessary to introduce guard bands between the actual (information-bearing) channels in order to avoid overlap and consequent adjacent channel interference.
2. There is essentially no inherent protection against multipath issues.
3. Much precious radio spectrum is occupied.

The last two aspects are of considerable significance and COFDM has been developed largely to overcome these two problems [1]. With this scheme, the signal is spread over thousands of narrow (FDM) channels, each of which operate at relatively low information rates. Each of the channels is coded and those portions of the signal that are to be transmitted through specific channels have the appropriate code attached such that no signal elements can become lost. In this way, the receiver can efficiently identify and reject undesired reflected (multipath) signals and only enable the strongest desired signal to be processed further through the TV receiver.

This COFDM approach can be used with the DVB standard and the bandwidth required is typically around 7 MHz.

A block diagram of typical COFDM DVB receiver RF and digital processing functions is shown in Figure 5.2. Not shown in Figure 5.2 is the necessary low-noise block (LNB) downconverter that is located in the aperture of the receiving antenna which is usually positioned outdoors. This LNB converts the incoming microwave (K or Ka-band) signal to ultrahigh frequency (UHF) (such as 1.8 GHz), which much more manageable using the subsequent electronics. At the RF input, the signal is downconverted further (forming IF1) ahead of the mixer, which forms the IF2 output.

Automatic gain control (AGC) ensures that the signal strength is maintained approximately constant where necessary. The final block is the COFDM demodulator chip that takes the IF2 signal input and delivers the MPEG2 or MPEG4 digital transport stream as output to the remainder of the receiver.

Facilities are also provided for the I/Q (in-phase/quadrature) data signals originating from satellite transmission or from the cable TV demodulator to control the digital processing chip.

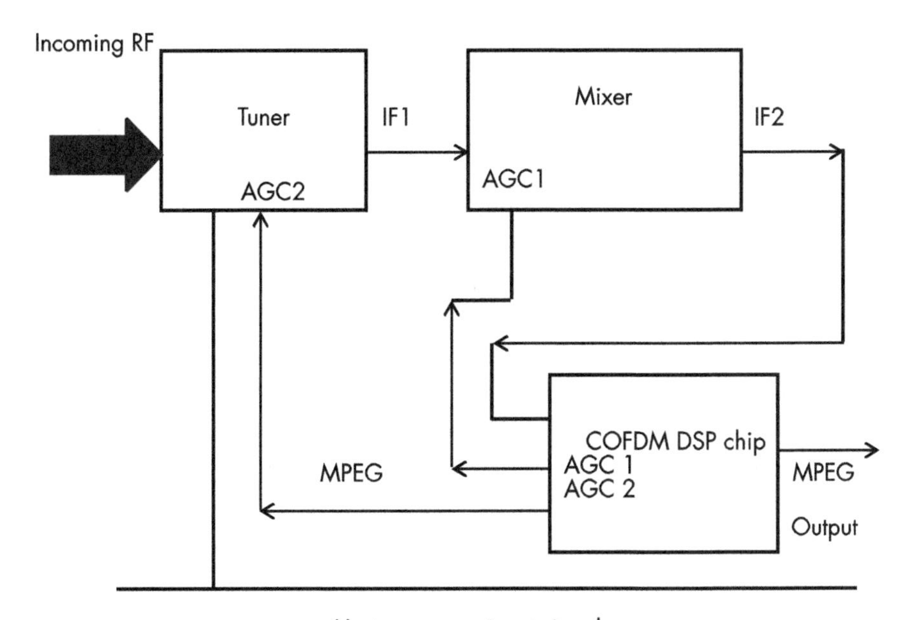

Figure 5.2 Block diagram showing a COFDM receiver (the COFDM DSP chip also requires I/Q data from the source transmitter).

5.5 SATELLITE TELEVISION

Multitudes of dish reflectors festooned around apartment blocks are testament to a generation using satellite TV. In most instances, it would make economic sense to install CATV serving groups of subscribers, but individuality combined with staggered installation timing has led to this rather ugly environment.

But, SatTV looks likely to be facing an uncertain future. In the 2020s, it looked like maybe the global SatTV market was headed for around $100 billion per annum. However, under pressure from web-based TV, CATV, the overall (global) total addressable market (TAM) actually fell by just over 1% by 2023.

Several factors tend to be driving this decline: internet (web-connected) TV, which can use cable (fiber or copper; see Chapter 6) and mmWave terrestrial and HAPS connections (see Chapter 8).

It looks likely that there could be a gradual shift to LEO-based SatTV receivers (LEO constellations are covered in Chapter 9). It also looks probable that LEO constellations will be providing direct-to-handset communications [2].

5.6 TELEVISION SCREENS

5.6.1 Resolution

In terms of resolution, TV screens are characterized by a specific term known as "nk" in which "n" is a numeric typically 4 or 8. The "k" quantity takes on the conventional abbreviation for 1,000 and practical TV receiver screens can typically be specified as 4k or 8k. A 4k screen is composed of 3,840 pixels horizontal by 2,160 pixels vertical, making a total of almost 8.3 million pixels. However, an ultrahigh-definition (UHD) 8k screen contains 7,680 pixels horizontal by 4,320 pixels vertical, making a total of almost 33.2 million pixels. Also, each pixel is so tiny that the viewer cannot identify it even when facing it close-up.

Watching almost any sports event, using an 8k TV is literally very much like being actually present as a spectator at the live event.

The progression of TV broadcast standards and required channel bit rates is shown in Table 5.2.

Another factor is that doubling the frame rate to 120 kHz increases the channel rate still further.

Table 5.2
TV Broadcast Standards and Resulting Channel Bit Rates

TV Broadcast Standard	Channel Bit Rate (Minimal Megabits per Second)
Basic	2.5
4k	25
8k	72

In the case of a multidwelling apartment block with high tens through hundreds of CATV subscribers (DVB-C; see Chapter 6), the main feed bit rate will amount to high hundreds of megabits per second or even several gigabits per second.

5.6.2 Twenty-First-Century Digital TV Receivers

In terms of digital TV, the SOCTA (twenty-first-century State of the Commercial and Technological Art) is so advanced it requires this separate section to even begin to adequately describe the situation. Yet it is essential for all of us to grasp the fundamentals, because only then can the demands on connection bandwidth be understood.

First, recollect that the vital importance of a handheld remote control unit became fully appreciated in the 1980s, the analog era. This aspect is even more important now that we are all well into the digital age with vast numbers of channels and apps, even if I often get really muddled over what button or sequence of buttons really does what on that remote.

Next we must consider the main display technologies now used, together with the implications for the broadband technology on which the typical twenty-first century digital TV depends.

5.6.3 Twenty-First-Century Digital TV Display Technologies

Some readers (or at least their parents or grandparents) will remember the age when displays, TV or computer, comprised evacuated cathode-ray tubes (CRTs) where the screen's back surface was scanned by a beam of electrons emerging from an electron gun that impinged on the back surface of the evacuated screen. The next time you tune in to watch an old film (certainly before 1990), look out for the TVs and computer monitors and observe how bulky these were in all respects.

They were also really heavy compared with modern displays because they required magnets to control the electron beams. Clearly, there is no comparison with the modern flat-panel displays, and Figure 5.3 illustrates the history of basic display technologies.

In all instances, each picture element is called a pixel, a vital concept first encountered in Section 5.6.1.

We are now concerned with the following two (distinct) types of flat-panel display technologies: liquid-crystal displays (LCDs), and organic light-emitting diodes.

For either of these technologies, characteristic faults include when a pixel is always off (i.e., permanently black) or, alternatively, when a pixel is permanently on (i.e., permanently white).

Most display applications will tolerate a few such faults, but when the number of faults approaches 10 or more, the screen is generally into gross and unacceptable failure.

5.6.3.1 Liquid Crystal Displays

LCD screens are almost certainly the most commonly encountered whether one is considering a monitor for a computer, the screen for a cell phone, or indeed a TV screen.

The first important point to note is that LCDs require a backlight that forms the origin of light to be either blocked or transmitted through the LCD crystals. The voltage stimulus changes the liquid crystal polarization rotations that modulate the brightness of the backlight passing through.

The varying voltage originates from the depression of a key on your computer keyboard (which is exactly what is happening as I type this information into my computer), or from appropriate signals associated with your smartphone, or information decoded from data streams entering your digital TV.

There are several million LCD pixels on a typical flat-panel LCD computer screen. Every individual pixel is driven by a digital signal

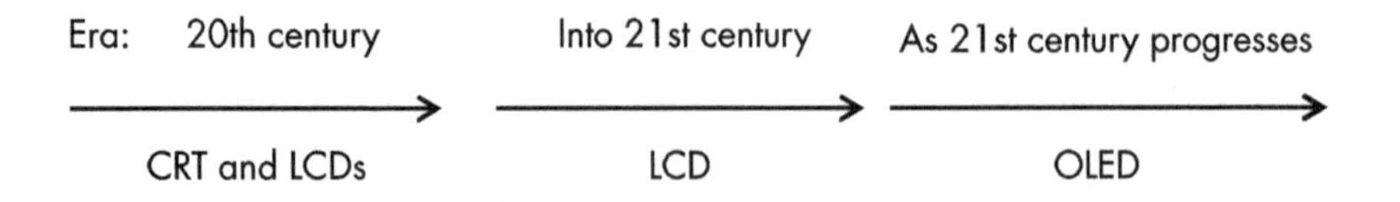

Figure 5.3 The evolution of electronic display technologies.

(causing a variable voltage) that has the information to determine the brightness of that pixel and the color of that particular pixel.

LCD screens have been commercially available since at least the late 1980s. Major manufacturers include Hitachi and Toshiba.

5.6.3.2 Organic Light-Emitting Diodes

Organic light-emitting diodes (OLEDs) are relatively recent contenders for the flat-panel stakes, mainly regarding digital TV screens. The operation of each individual OLED pixel is very complex in terms of semiconductor theory and operation and so only an outline is given here.

The key concept, which distinguishes OLEDs from LCDs, is that OLEDs are photoemissive diodes. The basic cross-sectional structure is shown schematically in Figure 5.4.

For operation, a voltage is applied across the OLED such that the anode is positive with respect to the cathode (akin to the familiar polarity requirement for batteries). Materials for anodes are selected for their optical transparency, electrical conductivity, and chemical stability. An electrical current flows through the device from cathode to anode and during this current's transit from the conductive layer to the cathode photo-emission occurs. This process is fundamentally due to semiconductor recombination that occurs under the electrostatic forces present in this region. A key feature is that in organic semiconductors holes (+) are more mobile than electrons which leads to a decrease in the energy levels of the electrons, which, in turn, leads to the emission of radiation, the frequency (or wavelength) of which lies in the visible ranges.

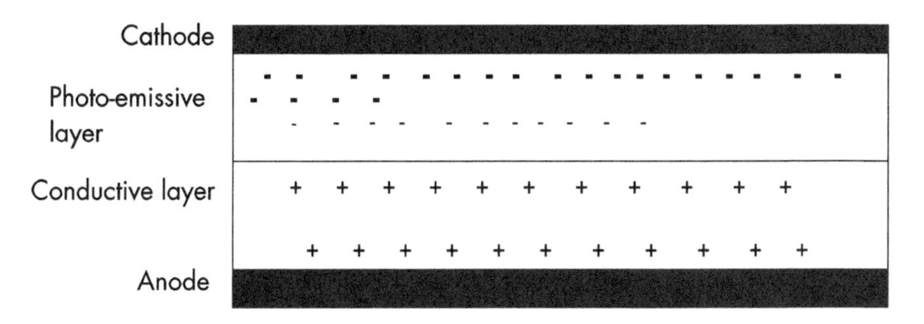

Figure 5.4 Schematic cross-sectional view of a bilayer OLED.

By driving each OLED pixel appropriately for the required intensity and color, the screen picture is rapidly generated. There exists a very substantial body of information concerning OLED pixels [3], but for present purposes the above brief treatment will suffice.

5.6.3.3 Complete Solid-State Television Screens

In the late twentieth century and the early years of the twenty-first century, most TV screens or computer monitors had square formats but this situation steadily changed so that an aspect ratio of 1.75 became the de facto standard. This means that most present-day displays have horizontal dimensions 1.75 times that of the vertical, whether LCD-based or OLED-based. A fairly typical screen format is shown in Figure 5.5.

In Figure 5.5, the side dimensions are given in millimeters because this provides a fairly good indication relating to the sizes of typical pixels and these data can be used here to calculate the approximate total numbers of pixels. However, the de facto standard diagonal sizes tend to be provided as diagonal dimensions in inches. In Figure 5.5, the diagonal is 1,800 mm, almost 71 inches, but there are also screens with 48 inches, 55 inches, 97 inches, and larger diagonals.

Assuming that each pixel is contained within each square millimeter of a screen, it is instructive to simply calculate the number of available 1-mm squares in a typical screen. Using the data in Figure 5.5, this calculation yields: $900 \times 1{,}575 = 1{,}417{,}500$ (i.e., just over 1.4 million pixels). In practice, screens comprise several million pixels,

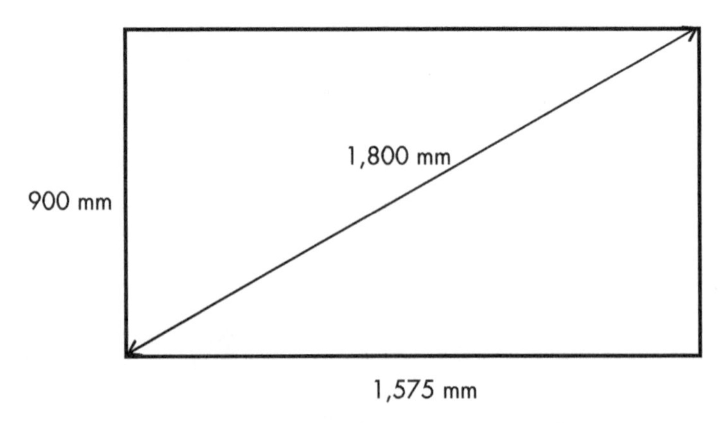

Figure 5.5 Schematic front outline of a fairly typical twenty-first-century TV screen.

and typical state-of-the-art (2020s) product specifications include "4k smart OLED TVs with 55-inch screens," attracting unit prices around $2,000. At the high end, it is possible to purchase a 97-inch 8k OLED TV comprising 33 million pixels (see Section 5.6.1), but, as should be expected, the unit price for such a product is more like $10,000. Manufacturers include Panasonic (Japan), LG (Korean), and Samsung (Korean).

As a note of caution, experience has shown that OLED pixels can be subject to burnout or image retention.

To avoid such issues, it is recommended that OLED TVs should not be left on standby for prolonged periods.

5.7 MODULATION SCHEMES

5.7.1 Choice of Modulation Scheme

The choice of modulation scheme is always important in communications, and DVB is no exception. Several schemes have been implemented, in particular:

- Binary phase-shift keying (BPSK);
- Quadrature phase-shift keying (QPSK);
- nQAM (e.g., 256QAM);
- Trellis-coded quaternary phase-shift keying (TC8PSK).

BPSK and QPSK are well known and fundamental types of digital modulation. QAM is described elsewhere in this book and the spectral efficiencies associated with each scheme (except TC8PSK) are well known; see Chapter 2.

5.7.2 Trellis-Coded Quaternary Phase-Shift Keying

Trellis encoding involves adding the coding information to the conventional modulation scheme (e.g., QPSK (i.e., four-level PSK)) to result in a signal characterized by successive values combining the coding itself with the conventional modulation scheme. A reverse tracking approach is employed whereby the previously received signal is compared with the more recent signal. In this way, a scheme such as TC8PSK has a greater tolerance to noise than either BPSK or QPSK.

Adopting TC8PSK enables the information transmission capacity to be one-third higher than available with the other, more traditional schemes within the same bandwidth. It is also necessary for a trellis-coded communications channel to be quasilinear in a similar manner to that associated with a vestigial sideband (VSB) analog local broadcast channel.

The remaining comments assume the signal is propagated through the troposphere. Unfortunately, the susceptibility to rainfall is increased, but techniques are available to counteract this drawback to some extent. Using TC8PSK information bit rates as high as 50 Mbps can be handled with associated received carrier-to-noise power ratios of about 10.5 dB (over a 27-MHz bandwidth), provided that the rain attenuation is not much above 4 dB. QPSK, the next best scheme, is limited to 40 Mbps, although 7 dB of rainfall attenuation can be tolerated at this bit rate. The 40-Mbps limit is too low for the 8k standard specification (see Table 5.1) and TC8PSK and QPSK are rarely adopted nowadays for DVB applications.

5.7.3 Quadrature Amplitude Modulation

Quadrature amplitude modulation (QAM) was described in Chapter 3 where it should be clear there are fundamentally many possible levels for this type of modulation, up to many thousands of levels.

For DVB-T and DVB-S, 256 QAM is usually applied, while for CATV (DVB-C; Chapter 6), levels as large as 16,384 QAM are even used in some instances. In general, the use of QAM means similar digital signal processing chipsets can be implemented into a wide range of types of systems including most of those described in Chapters 8 and 9 of this book.

5.8 DELIVERY SYSTEMS: LOW-NOISE BLOCKS, FEEDERS, AND COMPACT ANTENNAS

As shown in Figure 5.2, the RF input signal is generally taken directly from the antenna into the tuner and processed to form IF1, IF2, and then the MPEG2 digital signal processing. In most instances, this provides for effective reception.

Digital TV differs from its analog counterpart in more ways than just the technology. The quality of the antenna, its reception characteristics, and its immediately local electromagnetic environment are all of great importance. Reception critically depends upon the quality of the antenna, which is usually installed outside and therefore subject to all weather conditions. High-quality antennas are recommended in order to guarantee excellent reception in all but the most forbidding environments. In all instances, the pointing direction is critical for maximum signal reception.

In areas where the signal strength is relatively low, an LNA can be interposed between the antenna and the tuner to boost the signal strength. Ideally, the LNA should be positioned as near to the antenna as possible because only then is the noise figure maintained low.

With digital transmission, noise is much less of a problem than with analog and, therefore, such an LNA could usefully be integrated within the antenna input side of the tuner. However, DVB subscribers are really in an all-or-nothing situation because they are either in the position of having exceptionally high-quality reception or they have absolutely no picture and most probably no sound either.

Low-noise blocks (LNBs), more precisely called low-noise block downconverters, are required when the system is satellite rather than terrestrial. In this case, the familiar dish reflector is the conventional form of receiving antenna and the LNB is almost always located at the focus of the dish. Many suppliers offer products with the dish, the LNB, and the final downlink coaxial cable feeder connection integrated within a single unit.

The satellite signal, at a Ku- or Ka-band microwave frequency, is picked up by the dish antenna and concentrated by this parabolic reflector, being focused upon the LNB input horn. The first stage in the LNB is an LNA that amplifies this signal. Further signal processing within the LNB then downconverts the signal into the first IF. This is basically similar to the tuner function described above for the DVB receiver.

More compact planar antennas implement technologies such as microstrip arrays. These approaches dispense with the need for the bulky dish reflector and instead provide much more compact and environmentally-friendly products. Since the gain (i.e., the focusing capability) of such compact arrays is generally substantially lower than the values possible with dish reflectors, it is desirable for the satellite transponders to have higher values of microwave (or mmWave) output power. Reception using planar antennas can be compromised due to electromagnetic effects associated with the local environment, which means that engineers and technicians often prefer dish reflector designs.

For further related technology issues, refer to the final sections of Chapters 1 and 2, as well as [4].

References

[1] https://www.techtarget.com/searchnetworking/definition/COFDM.

[2] https://interactive.satellitetoday.com/via/october-2023/the-future-of-smartphones-with-satellite-direct-to-handset/.

[3] www.expertreviews.co.uk/tvs/1416142/what-is-oled.

[4] Edwards, T., *Technologies for RF Systems*, Norwood, MA: Artech House, 2018.

6

CABLE TELEVISION AND SATELLITE MASTER ANTENNA TELEVISION FEEDS

6.1 INTRODUCTION

Cable television has a long history, certainly in the developed world, and it continues to be important today in many towns and cities. Evolving standards apply and the emergence of relatively new TV technologies such as large-screen OLEDs (Chapter 5) drive bandwidth requirements upwards. Satellite master antenna TV (SMATV) is also covered in this chapter, although it is not as important as CATV. As defined in Chapter 5 under DVB, the appropriate descriptions here are DVB-C (cable) and DVB-S (satellite: SMATV).

6.2 CABLE TELEVISION SUBSCRIBERS AND DELIVERY TECHNOLOGIES

6.2.1 Subscribers

Driven by the ever-extending quality and quantity of available programs in many areas of life interest, CATV and SMATV continue to grow globally. CATV can be construed to stand for either cable or community television, but SMATV is universally understood to mean satellite master antenna television, sometimes called satellite distributed TV (SDTV), in the United States. According to most estimates, by around 2023 (when this book was being written), the global total CATV subscribers amounted to approximately 1 billion. By 2030, it has been forecasted that this will have more than doubled to exceed 2 billion subscribers. The details are shown in Figure 6.1.

Most subscribers are located in Europe, the United States, and Japan, but China, India, and South Korea are also increasingly important.

6.2.2 Cable Television Delivery Technologies

CATV is defined by the delivery of TV programs direct to customer premises using cable (cable TV), either copper cabling or using fiber optics, and there are basically three important types of configurations:

1. Existing CATV networks using copper cabling;
2. Existing (and some new) networks using combinations of copper and fiber cabling (hybrid fiber/coax (HFC));
3. New networks implementing fiber optics exclusively.

Option 1 and the already existing option 2 essentially comprise legacy networks, and it is important to appreciate that these remain very important commercially, mainly because there exist large numbers of such networks in many cities globally.

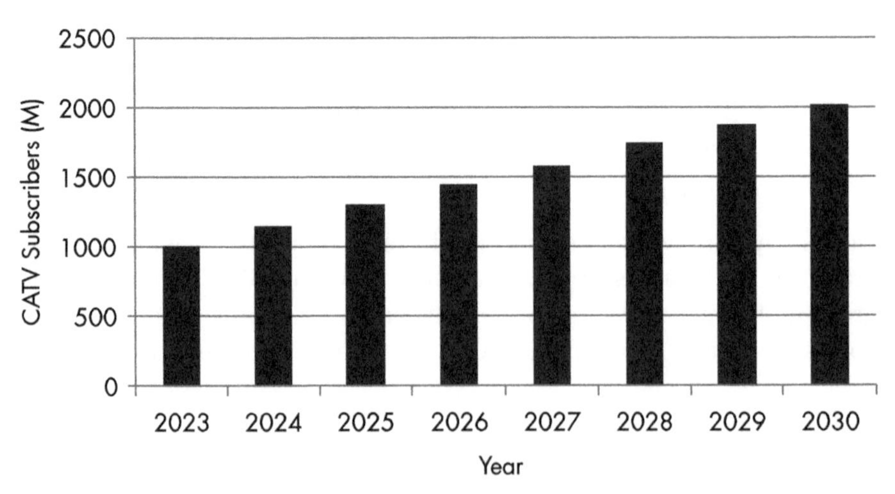

Figure 6.1 Global numbers of CATV subscribers (data generated by the author).

A typical HFC CATV network is shown in Figure 6.2. In Figure 6.2, coaxial cable connections are clearly visible as broken-line

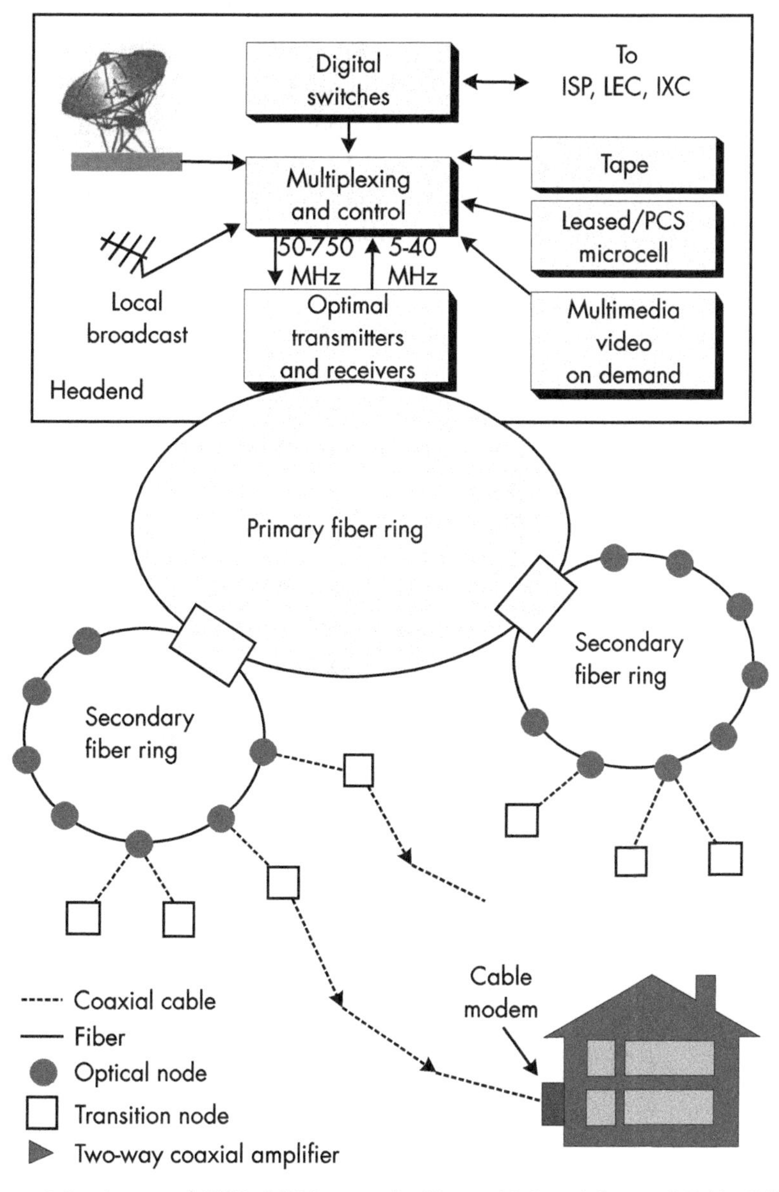

Figure 6.2 A typical HFC CATV network. (*From:* Nellist, J.G., and E.M. Gilbert, *Understanding Modern Telecommunications and the Information Superhighway,* Norwood, MA: Artech House, 1999.)

symbols on lines connecting to homes, and this is a very important feature because it represents many existing CATV networks.

6.2.3 Data Over Cable Service Interface Specification

In the later years of the twentieth century, it became clear that an international standard was needed to stretch the bandwidth of legacy copper cabling used by many CATV operators. To meet this need, in March 1997, CableLabs led a group of 15 other companies to introduce a new standard called Data Over Cable Service Interface Specification (DOCSIS). Among the 15 other companies, some of the larger companies included Cisco Systems, Intel, Motorola, Time Warner Cable, and Texas Instruments.

It is vital to appreciate that DOCSIS standards all distinguish between downstream and upstream CATV connections, with downstream referring to data flowing from the network to the final subscriber while upstream means the opposite (i.e., data transmitted by the subscriber up into the network).

The maximum operating frequency for CATV coaxial cable (DOCSIS standard) is, at the time of this writing, 1.8 GHz, and the current version of the standard is called DOCSIS 4.0, which allows the maximum data transfer speeds specified in Table 6.1.

Alternatively, under DOCSIS 4.0, the maximum throughputs (data rates) can be symmetrical (i.e., 5 Gbps in both directions).

As already stated, the original DOCSIS standard was introduced in 1997. DOCSIS 4.0 remains current at the time of this writing, and there were several intermediate versions introduced between 1997 and 2017. The 1997 standard provided for 40 Mbps downstream and 10 Mbps upstream (note megabits per second), and 5 years elapsed before the upstream rate increased to 30 Mbps. Several years later, the world raced into the gigabit era, which led to today's gigabit per second data rates.

Table 6.1
Data Transfer Speeds Under DOCSIS 4.0

Direction of Data Flow	Maximum Data Rates (Gbps)
Downstream	10
Upstream	6

There is also a EuroDOCSIS version predicated by the need for the standard to accommodate the differences between European and North American channel frequency bandwidths. These differences are based on the upper operating frequencies in the RF channel bandwidths as shown in Table 6.2.

Note that Europe's Phase Alternating Line (PAL) may not necessarily be all that friendly, while according to some mischievous people, North America's standard stands for Never The Same Color.

All DOCSIS standards are fully supported by the ITU and the IEEE, for example, (ITU) b-ITU-T J.224 (which includes XG-PON fiber-optic); see Chapter 3.

Regarding modulation schemes almost all modern CATV systems use QAM (see Chapter 2). For downstream, the modulation level can be as high as 16,384 QAM, while for the upstream signal the maximum is more like 4,096 QAM. Such high-level modulation schemes would be highly subject to external noise and interference for atmospheric or space-based systems (Chapters 4, 7, and 8) but can be accommodated for CATV because the networks are electrically (and optically) enclosed.

Nowadays the dominant delivery technology is fiber optic, specifically PONs, covered in Chapter 3.

6.2.4 Cable Television Distribution Via Passive Optical Networks

PONs were introduced and described in Chapter 3 where Section 3.3.3.4 focused on applications to CATV. In that section, it was pointed out that standards known as Gigabit Ethernet PON (GPON) and Ethernet PON (EPON) are increasingly important.

GPON provides three layer-2 networks: ATM for voice, Ethernet for data, and proprietary encapsulation for voice. GPON also provides 1.25 Gbps or 2.5 Gbps downstream and upstream bandwidths scal-

Table 6.2
DOCSIS and EuroDOCSIS

Region	TV Standard	RF Channel BW (MHz)
Europe	PAL/DVB-C	8
North America	NTSC/ATSC	6

able from 155 Mbps to 2.5 Gbps. One commercially available example is Ericsson's BLM 1500 product.

EPON uses Ethernet packets instead of ATM cells. It employs a single layer-2 network that uses IP to carry data, voice, and video. It usually offers 1-Gbps symmetrical bandwidth, making it particularly popular in modern networks.

However, the main rationale for implementing PON into CATV networks is the use of DWDM. A wide range of standards have been introduced by the ITU, related to the application of DWDM for GPON. These standards are GPON (or EPON), 10G PON (or 10G EPON), 25G PON, 50G PON, and XGS-PON.

Each PON standard is assigned a specific wavelength according to ITU PON wavelength plans, which are shown in Figure 6.3.

The wavelength plans in Figure 6.3 provide many more specification standards than just CATV and are grouped into two regimes as follows:

- GPON, 10G PON, TWDM-PON, and RF CATV;
- 50G PON.

Throughout the chart of Figure 6.3, DS means downstream and US means upstream. TWDM-PON refers to time-WDM-PON, which provides for upwards of four more wavelengths per fiber, each capable of delivering NG-PON2. This specification was first introduced by Nokia in 2014 [1].

Applying to 50G PON, UW means ultrawide (band): UW1 covering 1,290–1,310 nm and UW2 encompasses 1,260–1,280 nm. An exception is the narrowband option within UW1, which is centered on 1,300 nm with a symmetrical bandwidth of 4 nm (± 2 nm) (i.e., a 0.3% overall bandwidth).

CATV occupies a 10-nm band ranging from 1,550 to 1,560 nm.

There are some other features that are relevant to CATV but not shown in Figure 6.3. Among these, RF over glass (RFoG) is an important standard in the context of CATV and the two bands specified are as follows:

- Upstream (US) (within PtP WDM RFoG): 1,290–1,330 nm;
- Downstream (DS) (within PtP WDM RFoG): 1,603–1,625 nm and downstream (DS) (RF Overlay/DOCSIS): 1,546–1,554 nm.

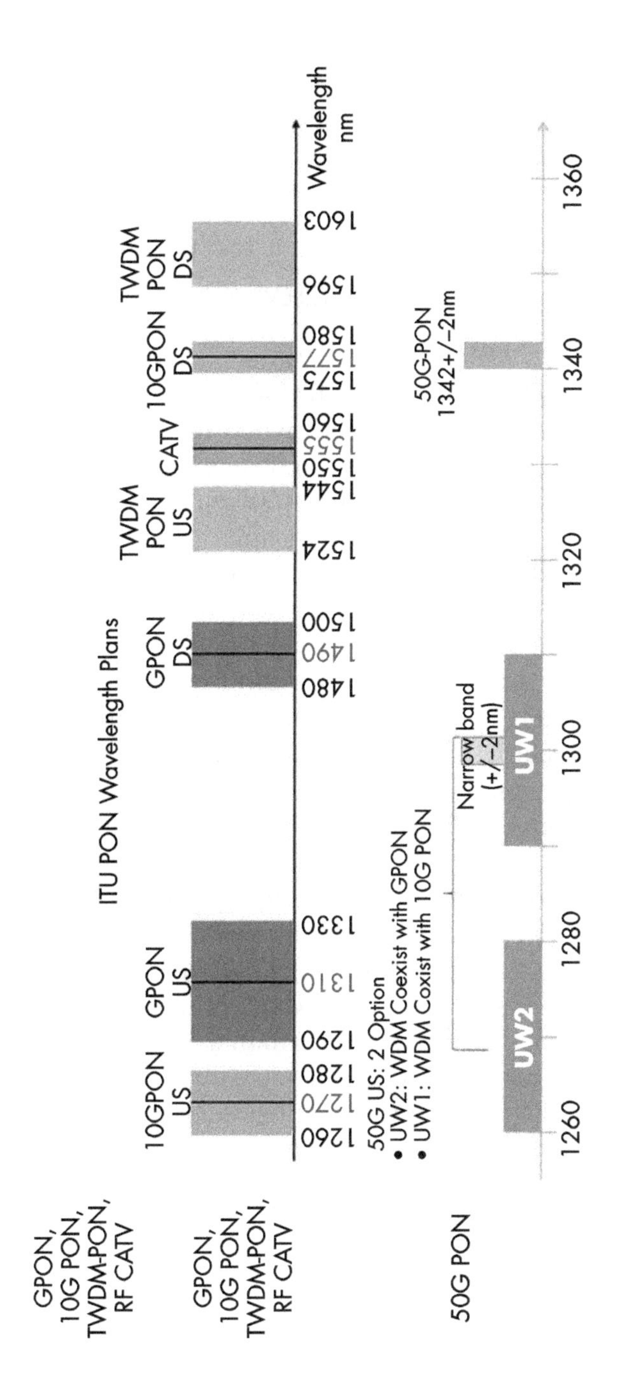

Figure 6.3 ITU PON wavelength allocations.

This is fully compatible with the DOCSIS standard described in Section 6.2.3.

Geoff Varrall's book [2] provides substantial information with respect to these standards (also much on digital TV) and the article by Zhang et al. [3] includes more detail regarding the standardization process.

6.3 SATELLITE MASTER ANTENNA TELEVISION

6.3.1 Block Dwellers

The types of CATV networks described above work well enough and provide good returns on investment when they apply to neighborhoods with many individual homes and small apartment blocks having perhaps up to around 10 or so units. When large apartment blocks with hundreds or even thousands of units (usually termed multi-dwelling units (MDUs)) or big hotels are considered, then it is usually not cost-effective to operate CATV networks as described above.

Instead, the block or hotel owner will often be attracted to turn the operation into essentially its own CATV network. In such cases, certainly with hotels, the concept of community hardly applies and another term is required to define this approach. Because many of these types of systems have traditionally been exclusively satellite-fed, with the antenna and LNB naturally rooftop-mounted, the name satellite master antenna TV (SMATV) was coined. An equivalent term, satellite distributed TV (SDTV), is frequently used in North America.

A schematic physical illustration of SMATV is given in Figure 6.4. The signal enters the satellite antenna, the master antenna, and is transferred to the head end via a coaxial cable feed. From the head end, the entire block is fed with distributed TV channels. This distribution network may totally comprise coaxial cables (common), or very often a mixture of coaxial and fiber cabling (HFC; see Section 6.2.2), or entirely fiber (PON), particularly in new installations.

Not shown in Figure 6.4 is the control center for the system that is usually located in the administrative staff offices in the case of a hotel. Individual hotel rooms have the usual vast array of channels, most of which will be free (i.e., combined with the price of the room), while programs on several other channels will require specific ordering, normally online. Although often extending to several hundred

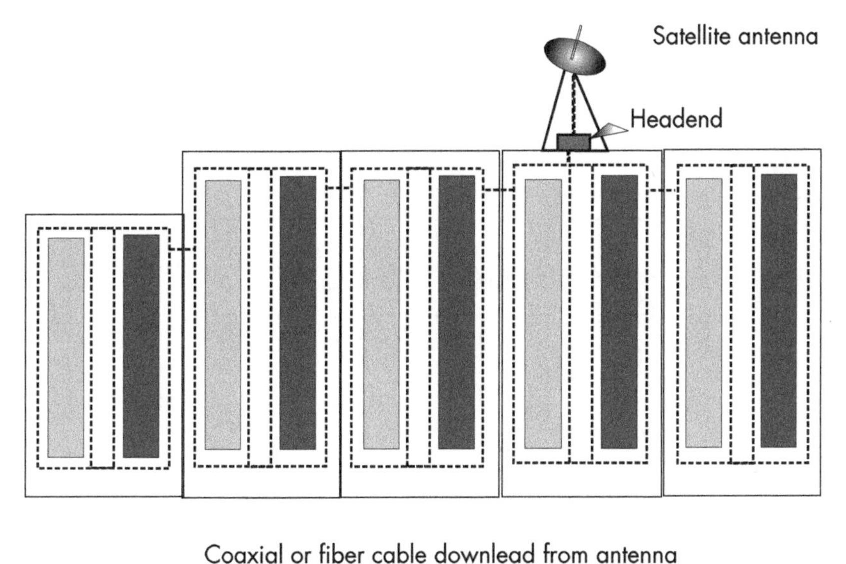

Figure 6.4 SMATV in a typical hotel complex. *From:* [4].

meters, the distances to be spanned in large apartment blocks or hotels are generally much shorter than in city neighborhoods, so, until recently, fiber-optic transmission was rarely implemented in these installations and coaxial cable systems (or HFC) predominated.

With the tangible advantages of freedom from electrical interference and capability for high-speed transmission, optical systems (PON) are becoming much more commonplace in MDUs at least. An example of a typical SDTV system topology is provided in Figure 6.5 in which solid curved interconnects represent coaxial cables and wavy line fills indicate optical fiber cables.

As usual, the signal first enters the satellite antenna and is then transferred along a coaxial cable downlead to the next unit. This first major unit is a fiber-optic transmitter capable of feeding from 8 to 16 optical output channels to possibly hundreds of nodes in the complex. The fiber-optic transmitter accepts the broadband electrical signal input and converts it into the multi-output series of optical signals. Serving each node, there is an SDTV fiber-optic receiver that feeds each apartment with the required signal.

Within each apartment, a conventional coaxial cable RF input connects to the TV receiver, including full internet access. Set-top

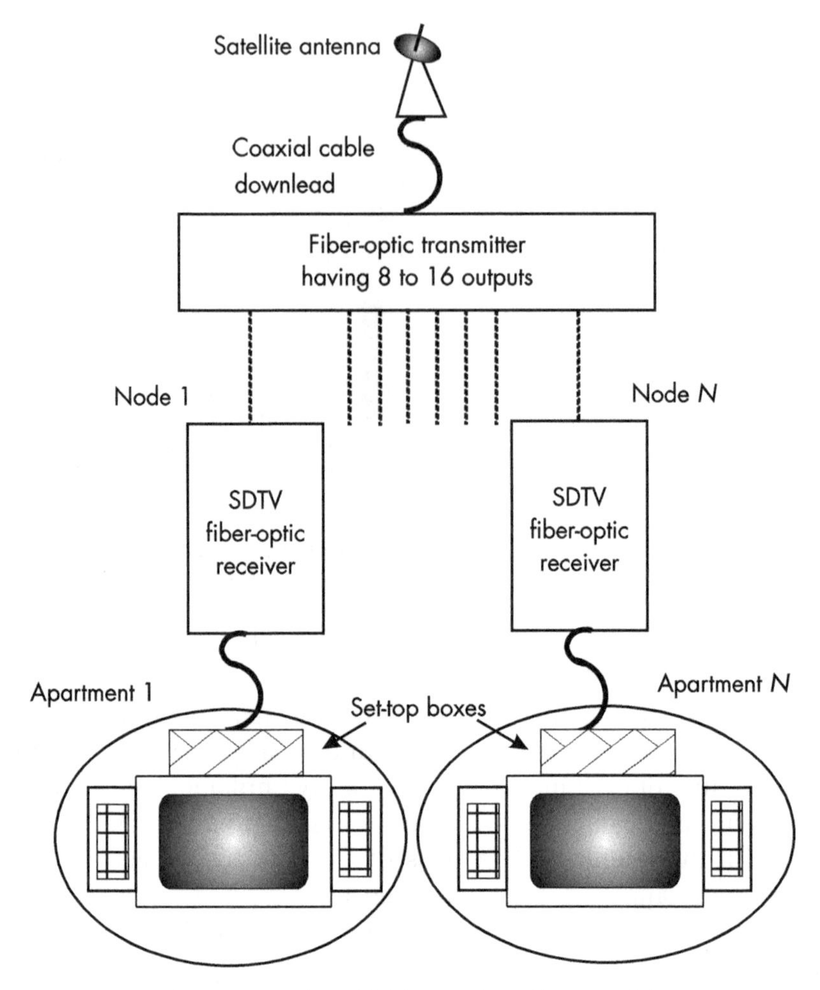

Figure 6.5 Satellite distributed TV SDTV (using DVB-S). *From:* [4].

boxes are indicated in Figure 6.5, although most modern digital TVs embody all the required electronics internally. Such a system provides the advantages of fiber distribution (PON) in an economical manner.

Incidentally, the programming demands will differ markedly where hotels are concerned compared with the viewer profile associated with domestic units. Because large hotels generally cater to both national and international travelers, appropriate programs are required to suit these guests. Certainly, local events, such as news and highly localized sports, will usually not be attractive to international hotel guests. These factors alone are powerful drivers for the hotel owners to introduce and run their own SMATV networks.

6.3.2 Community or "Just You"

One key determinant in deciding upon the configuration of cable feeds is whether it is to serve a multitude of people or just an individual person or family. For multichannel TV reception to individual home dwellers or those living in relatively small MDUs, at least there is almost always the option of having individual satellite TV facilities. These traditionally provide receive-only access to typically a similar number of similar channels to those available from a CATV operator. Also, there is the independence from such an operator because it is easy to consider removing an existing installation and simply replacing it with a new one, perhaps obtained from a completely different store or online mail order company such as Amazon. Even if the installation is not changed, there is always the opportunity of realigning the antenna to improve reception where this is deemed necessary.

Basically, with individual satellite TV facilities, there is the perception (real or imagined) of much greater freedom and control over the incoming TV programs. However, the prospect of having a vast population of satellite TV dish antennas littered across the outer walls of buildings such as MDUs is extremely unattractive and would almost certainly contravene local regulations and laws in many areas. Therefore, either the SMATV or the CATV approach is essential in these instances.

For individual or family units, CATV can automatically ensure that, with minimum effort, the latest programs and program reception technology are always available and that the subscribers are frequently reminded about this availability. With individual satellite TV, this is either leased through one of the large leasing companies or it is purchased outright. If leased, then the most up-to-date technology is immediately available from the owning company, so the situation is somewhat similar to that applying to CATV. If purchased direct, then the individual owners must pay for the cost of upgrades themselves.

6.3.3 Communications Technologies

Technologies available for communications systems (including standards) are constantly evolving and details are covered in this book.

Whether these are terrestrial, cabled, or otherwise, it is certain that new advances will become available over future years and decades.

References

[1] www.nokia.com/blog/twdm-pon-taking-fiber.new.wavelengths/.

[2] Varrall, G., *5G and Satellite RF and Optical Integration*, Norwood, MA: Artech House, 2023.

[3] Zhang, D., et al., "Progress of ITU-T Higher Speed Passive Optical Network (50G-PON) Standardization," *Journal of Optical Communications and Networking*, October 2020, pp. D99–D107.

[4] Edwards, T., *Gigahertz and Terahertz Technologies for Broadband Communications,* 1st ed., Norwood, MA: Artech House, 2000.

7

TELEPORTS

7.1 WHAT IS A TELEPORT?

A teleport is a gateway for satellite communications that is almost always visually characterized by a ground-based installation comprising several satellite receiving dishes and/or radomes covering such. The term "teleport" is short for "telecom-port."

In this chapter, we first describe the nature of teleports and follow this up with a brief description of some examples. The various types of teleports are defined, again with examples of specific instances.

Where exclusively corporate high-speed internet access is required, then very small aperture terminals (VSATs) adequately fit the need. Worldwide, the total number of VSAT sites has grown exponentially over the decades since around 1980 to have reached several million at the time of this writing and may well approach some hundreds of millions by around 2040. Although the RF technology will likely change radically, implementing many more flat-panel antenna arrays, there will be an increasing free-space optics element.

Teleports are much more sophisticated than VSATs, being comprehensive, shared-use, and most often city-based telecommunications facilities. They are generally:

- Advanced;
- Comprehensive;

- Capable of flexible, organic growth;
- Intelligent;
- Physically large (in most instances).

The beginning of the third millennium marked the third decade of the teleport—a concept that began in the 1980s with small shared-use satellite communication hubs that were being developed by North American entrepreneurs. This essentially followed the deregulation of satellite services in North America.

Today's teleports mostly comprise highly sophisticated developments, to which most of the examples described in Section 7.3 attest. Naturally, the explosive growth of the internet has further spurred expansion in teleport activity because teleports provide good opportunities for broadband multimedia connectivity. A broadly typical view of a teleport is shown in the upper picture of Figure 7.1 (the lower picture comprises a flat-panel antenna).

Prospects for the introduction of ground-based flat-panel antennas are discussed in Chapters 8 and 9. We now consider the basic forms of teleports before proceeding to describe a range of examples.

There are four basic forms of teleport and, working upward hierarchically, these are indicated conceptually in Figure 7.2.

Figure 7.1 Photograph of a typical teleport. (*From:* Figure 8.7 in [1], with permission.)

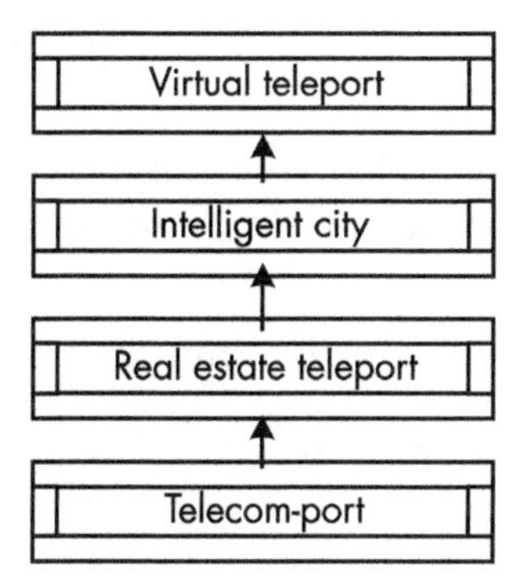

Figure 7.2 The four basic forms of teleport. *From:* [20].

Each of these forms of teleport is now described, working upward from the lowest in the hierarchy. The telecom-port format is characterized as follows:

- A telecommunications hub providing services on a profit basis;
- May be either independent or carrier-owned;
- An "information port" rather than an airport or a seaport;
- A facility providing for the shared use of complex and expensive facilities.

This is really the most basic form of teleport, that is, a port for telecommunications.

The basis is identical to that of any port including airports or seaports and amounts to the attractive economics of sharing the costs of a relatively comprehensive yet expensive facility.

Telecom-ports may be independent operations or they may be subsidiaries of major carriers or even other corporate parents. They usually combine both satellite and terrestrial (mostly fiber-optic) communications and they provide practically all the communication needs of any conceivable customer. This includes the most complex and bandwidth-hungry services imaginable and full motion videoconferencing is almost always provided.

The real estate teleport has the following general characteristics:

- It adds intelligent and fully networked commercial real estate to the telecom-port.

- It is usually developed under a public/private partnership agreement.
- It provides value-added services to its tenants including upgraded land upon which these tenants can build and expand their operations.
- There is an overall appeal to multinational companies.

The real estate teleport is an extension of the fundamental telecom-port concept, which is extended by means of an intelligent fully cabled network that is often a mix of coaxial cable and fiber optics (see also CATV in Chapter 6)—increasingly mainly fiber. This approach opens up new economic development opportunities and provides a secondary profit center for the local region in which this type of teleport has been built. Nearby organizations can promote it as an ideal facility for companies with extensive telecommunications needs—and that includes most reasonably large companies.

Given this raison d'être, an alliance between local government and business continues to be the major pattern for the development of most real estate teleports. Clearly, the businesses look for the profitable opportunities while the government agencies interests are principally those of engendering local economic development. These real estate teleports are often particularly attractive for second-world or even occasionally third-world economies because they provide world-class office and telecommunications facilities that help to pull in large multinational corporations. All the main Korean and Malaysian teleports are real estate types.

The intelligent city is the next level up in terms of sophistication and this is illustrated conceptually in Figure 7.3.

Clearly visible are microwave and millimeter-wave antennas for both satellite and terrestrial communications. Some of these provide base station/switching center interconnects for mobile (cellular) communications, while others are dedicated terrestrial communications links. Not visible, because they are deeply embedded, there are also extensive upgradable fiber-optic networks and networked computer systems. Also not visible are any free-space optical facilities, discussed later in this chapter.

In essence, this is either an entirely new urban center or the intensively redeveloped downtown region of a city. There are

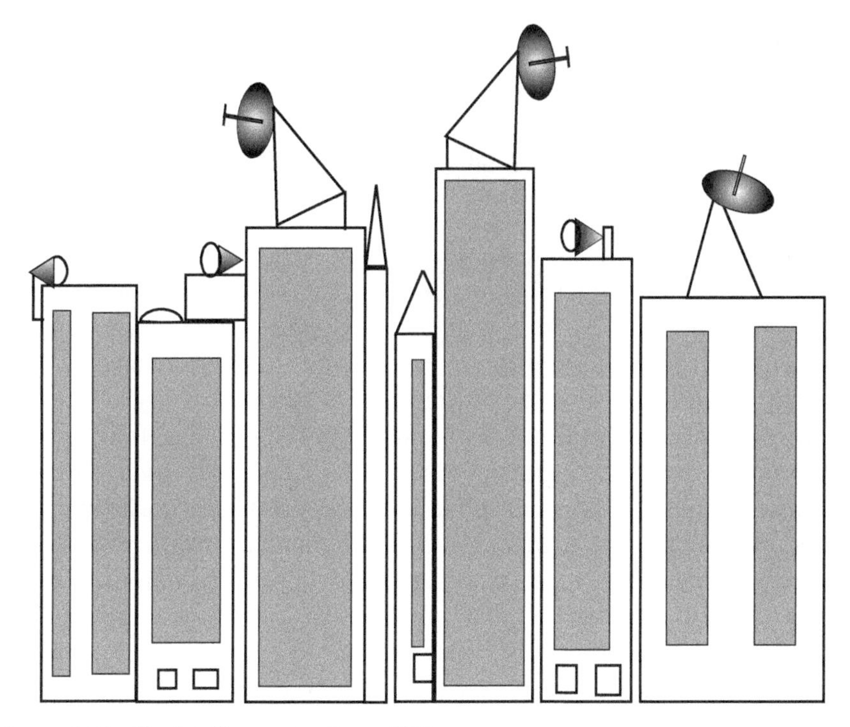

Figure 7.3 The intelligent city. *From:* [20].

state-of-the-art communications and information services made available to business, academe, government agencies, and other institutions both on-site and also off-site. All the systems are linked to a central teleport operations center. Because advanced systems are already in place, occupying companies can simply plug in their network components, and few, if any, enhancements are necessary.

This type of teleport is becoming extensively implemented in many economies, predominantly in Japan, some European cities (including Paris), and also the Rio de Janeiro teleport.

The virtual teleport is at the top level of the hierarchy shown in Figure 7.2 and it is also the most recent in terms of concept. This type of teleport virtually integrates existing advanced telecommunications infrastructures usually with some added value from introduced subsystems and services. Virtual teleports are:

- A cooperative venture among business corporations, academe, government agencies, and nonprofit organizations;

- An asset for infrastructure-rich communities that have no requirement for new telecommunications facilities or high-tech business parks;
- A one-stop shopping gateway to existing networks including satellite communications and associated services.

Those communities who are already rich in terms of their telecommunications infrastructures have no need for yet another new high-speed system. This is also an important consideration when markets are somewhat depressed, as during the 2008 recession. The virtual teleport provides for a pooling of resources between the types of organizations mentioned above and creates a virtual telecom and information hub by integrating assets owned by such diverse organizations. Although no formal, physical, teleport exists, individual and corporate users view it through their phones and multimedia computer terminals as though they were using a teleport. The key point is that, to establish a virtual teleport, no new telecommunications infrastructure is necessarily required. Instead, a teleport is evolved from the system of intelligent interconnects available from the existing infrastructure.

It is almost certain that virtual teleports will steadily gain ground as the third millennium progresses—mainly in already highly developed economies such as Korea, Japan, Singapore, the United States, and parts of Europe.

Across the entire teleport scene, for all types of installation, major international corporations have substantial stakes in teleports. The great majority of teleports comprise Intelligent Cities.

7.2 ROLES OF TELEPORTS

Much has already been described in Section 7.1 concerning the roles of teleports, and, summarizing, these roles include:

- Providing a highly efficient on-ramp for the internet and the information superhighway;
- Providing services on a profit-oriented basis;
- Providing for the shared use of complex and otherwise expensive facilities;

- Appealing to multinational corporations as well as to other types of users;
- Providing for a pooling of resources.

Teleports should not be seen as separate stand-alone entities but rather as installations or even virtual concepts that become naturally integrated with other surrounding and international systems and networks. This applies to essentially all forms of communications systems, including CATV and SMATV or SDTV (see Chapter 6). The concept also applies to primarily data networks such as LANs, metropolitan area networks (MANs), and wide area networks (WANs), and examples of the possible interconnections are shown in Figure 7.4.

Connections from SMATV or CATV networks are predominantly one-way whether between each other, between them and the teleport, or from CATV to MAN. All other interconnections, to and from LAN, MAN, or WAN and teleport, are two-way real time and most are provided by optical fiber. MAN interconnections are typically operating

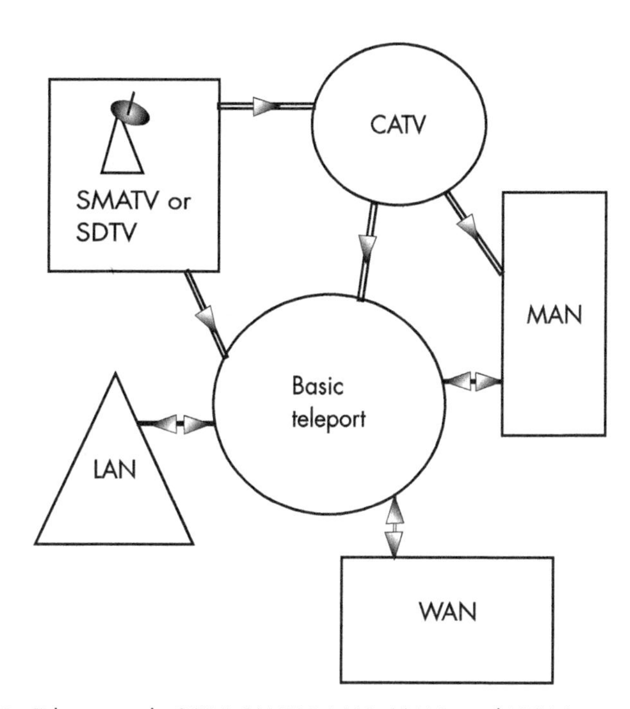

Figure 7.4 Teleport with CATV, SMATV, LAN, MAN, and WAN interconnections. *From:* [20].

at 155.52 Mbps (SDH level-1). A corporate LAN increasingly has multigigabit information rates, even hundreds of gigabits per second.

Teleports are highly flexible and can be transparent to the types of technology involved. In terms of services, it is of no consequence whether cabled or wireless interconnections are implemented. The specific type of technology applied at various nodes will be appropriate to the economic limitations, the geography and topology existing, and the information throughput required. Another very significant aspect is the capability for organic growth, because old systems or subsystems may be replaced by new and more appropriate subsystems as the requirements change and need upgrading.

Most important of all, once future-proofed networks are in place, the emphasis can shift toward software implementations of as many upgrades as possible. In short, as teleports are software-managed and controlled, they can be software-reconfigured and upgraded. In the future, a major trend will probably be for the physical growth of cities and other similar communities to slow, but such entities will expand constantly in terms of information and communications growth. In this respect, it is that teleports will have increasingly influential roles to play.

7.3 THE WORLD TELEPORT ASSOCIATION AND TWENTY-FIRST-CENTURY TELEPORTS

This is the main section of this chapter. In this section, the World Teleport Association (WTA) [2] is briefly introduced and a substantial range of representative teleports (and teleport operators) are described.

The WTA [2] (www.worldteleport.org) is an organization to which many (certainly not all) teleport owners and operators belong. The benefits of such membership include qualified tiers of membership levels and relatively easy communications with other members.

The WTA offers four tiers of membership ranging from the most comprehensive (Tier 4) down to the most basic (Tier 1). Because the WTA teleports described here all qualify as either Tier 3 or Tier 4, only the facilities and procedures associated with these two upper tiers are outlined here. These descriptions broadly follow the WTA published material, but minor adjustments have been made for consistency.

7.3.1 WTA Tier 3

The facilities for Tier 3 are as follows:

- There is limited redundancy with automated monitoring and switching for power, high-voltage AC, transmission systems and cabling, and full redundancy for terrestrial connectivity at up to MPLS OC-3 (optical) or equivalent capacity.
- There are no frequency restrictions in a long-term leased location with a low risk for natural disaster, where multiple antennas are needed for full arc coverage.
- Antennas are motorized, capable of tracking, and equipped with the highest-quality waveguides, dehydrators, and other systems.
- The facility has a moderate safety program and a moderate volume of reportable accidents.
- Physical access is well controlled by personnel, automated systems, and perimeter barriers. The teleport is located an acceptable distance from police, fire, and medical facilities and is staffed for part of each day, 7/365, and is accessible afterhours.
- Overall control is located remotely with redundant facilities, operates and is supported 24/7, and employs a high-quality monitor and control system. Testing, training, and related procedures assist in maintaining good vigilance against incidents.

The procedures for Tier 3 are as follows:

- The capacity plan is updated frequently, implemented with minimal delay, and often reviewed against standards.
- Service continuity plans are reviewed and tested often.
- An effective information security plan is often reviewed and circulated to employees, and system updates and audits are conducted frequently.
- An effective change management process involves some appropriate staff, is correlated to issues, and is reviewed often.
- An effective configuration management system records configurations for each service and ensures that configurations are fully implemented within the facility.

- Key performance indicators (KPIs) are automatically monitored, manually analyzed, undergo frequent review, are updated when standards are affected, and result in effective management of events, basic incidents, and major incidents.
- They are formally accredited or working in alignment with major international standards denoting evidence of the best practice across many areas.

7.3.2 Tier 4 (the Top Tier)

The facilities for Tier 4 are as follows:

- They have full redundancy with automatic monitoring and switching for power, high-voltage AC, transmission systems and cabling, geographic diversity, and terrestrial connectivity at MPLS OC-4 (optical) or higher capacity.
- There are no frequency restrictions at a company-owned location with a low risk for natural disaster, where most antennas have the full visibility of the GEO arc.
- Antennas are motorized, are capable of advanced tracking, and are equipped with the highest-quality waveguides, dehydrators, and other systems.
- The facility has a strong safety program and a low volume of reportable accidents.
- Physical access is thoroughly controlled by personnel, automated systems, and perimeter barriers. The teleport is located close to police, fire, and medical facilities and is staffed 24/7/365.
- Control is located on-site, operates and is supported 24/7, and employs an advanced monitor and control system. Testing, training, and procedures maintain high vigilance against incidents.

The procedures for Tier 4 are as follows:

- The capacity plan is updated frequently, implemented promptly, and frequently reviewed against SLAs.
- Service continuity plans are reviewed and tested frequently.

- A comprehensive information security plan is frequently reviewed and circulated to employees and involves frequent system updates and external audits.
- A comprehensive change management process involves all appropriate staff, is correlated to issues, and is frequently reviewed.
- A comprehensive configuration management system records configurations for each service and ensures that configurations are fully implemented within the facility.
- KPIs are automatically monitored and analyzed, undergo frequent review and improvement, and result in high-quality management of events, incidents, and major incidents. Documented evidence of processes, tools, and culture ensure the highest standard of event and problem management across the entire service operation.
- The procedure is formally accredited to a wide variety of international standards denoting evidence of a high degree of best-industry practice across all areas.

As examples of key differences notice how under facilities for Tier 3, OC-3 capacity is adequate, whereas for Tier 4, this must be OC-4 or higher. Also, under procedures for Tier 3, KPIs need to be manually monitored, whereas for Tier 4 this has to be automatically monitored. The reader is invited to observe further differences.

From this point onwards in this chapter, a substantial range of representative teleports (and teleport operators) is described, starting with the large Italian corporation known as Telespazio [3]. This concern operates internationally and is highly significant in terms of operating teleport businesses. Because of its importance, Telespazio is described next.

7.3.3 Telespazio

From its headquarters in Italy, the company known as Telespazio ranks among the world's leading satellite services operators. Telespazio's operations are based on a domestic and international network of space centers and teleports. In Italy, these comprise:

- The Fucino Space Centre located in the Abruzzo region (a WTA Tier 4 teleport). This is described in Section 7.3.11.1.
- The Lario, Matera, and Scanzano Space Centres.

The company also has activities within the Italian Defense Centres of Pratica di Mare and Vigna di Valle, Rome, and it supports the French Defense Centres of Creil and Maisons-Laffitte. Easterly within Europe, Telespazio operates its own teleport in Romania. In South America, Telespazio operates its own teleports in Argentina (Malargue, where it manages ESA's Deep Space 3 station) and also in Brazil. In Africa (Kenya, Malindi) where it supports the Italian Space Agency's Broglio Space Centre. Also, Telespazio has business with the Kourou Space Centre in French Guyana.

7.3.4 Japanese Teleports

In Japan, teleports have evolved substantially over the decades of the late twentieth and early twenty-first centuries. Teleports in the two main cities of Osaka and the capital Tokyo are next considered in some detail.

7.3.4.1 The Osaka-Based Teleport

This teleport is owned and operated by A-Cosmos Co. Ltd., Osaka, Japan [4]. A-Cosmos is 25 years old and is a WTA member. The main systems operate in Ku-band and the following applications are provided:

- Satellite news gathering (SNG);
- TV receive-only (TVRO);
- Ultrahigh definition TV (UHD).

A-Cosmos also has some microwave C-band uplinks positioned in Ibaraki and Yokohama.

7.3.4.2 Tokyo Teleport (Aruji)

Established in 1995, Aruji Co. Ltd. [5] began uplink operations by initially offering low-cost satellite broadcasting solutions to customers involved in digital broadcasting. Following this, the company then established their teleport center, which now includes various SNG vehicles, all with necessary licenses. This global teleport is connected to the company's Tokyo Teleport Aruji using a dedicated optical fiber cable. Aruji is not a WTA member.

Around two decades into the twenty-first century, Aruji's major achievements include:

- The Rugby World Cup in 2019;
- The G20 Osaka Summit;
- The UAE's rocket launch in 2020, initiating the Mars mission.

The fact Aruji, rather than the relatively local A-Cosmos (see Section 7.1.1), covered the G20 Osaka Summit is interesting. Further services provided by Aruji include:

- Production;
- Play out;
- Post production;
- Video shooting;
- Live broadcasting;
- Global hotel TV.

7.3.5 The Rio de Janeiro Teleport Center

The Italy-headquartered company Telespazio [3] owns and operates three teleports in its home country of Italy and three in Brazil (Rio de Janeiro, Itaborai, and Marica). The company also has one in Argentina and one in Romania.

Regarding the Rio de Janeiro Teleport Center:

- Telespazio has two teleports in the center of this city, known as RB1 and BT7.
- The two teleports offer satellite telecommunications, backhaul, monitoring, and control services in different markets as well as services in the area of geoinformation for data distribution to end users. The total land area occupied by the teleports is 150,000 square meters and there are 14 antennas.

5G connections are well covered and there is a gateway to the OneWeb LEO constellation (see Chapter 9 for details concerning OneWeb).

7.3.6 Teleports in Malaysia, Indonesia, and the Philippines

In the 1990s, Malaysia inaugurated a project called the Multimedia Super Corridor (MSC), which was an ambitious concept at the time. As a consequence of unforeseen economic pressures, this MSC project failed to get off the ground. Regional teleport developments have progressed and, for example, a company named Kakuchopurei operates several teleports located in various parts of this region [6]. Kakuchopurei also operates several Genshin Impact Teleport Waypoints dotted around the Philippines as well as Singapore.

PT Primacom Interbuana (Primacom) [7] is one of the major communication solution providers in Indonesia. Like many providers, PT Primacom Interbuana (Primacom) began its journey in the late twentieth century by providing connectivity services through a satellite communication system with a VSAT. The gateways are mainly Ku-band. Because the market continues to demand advanced communication network technology with greater efficiency, Primacom continues to expand its business line by offering services through its teleport such as fleet management, a data center, cloud, and internet services.

With the increasing bandwidth utilization through 14 transponders using 6 satellites, Primacom has covered almost 10,000 connection points for various industries, including banking and financial institutions, engineering procurement construction (EPC), telecommunication, oil and gas, mining, plantation, manufacturing, transportation and logistics, and maritime. PT Primacom is a WTA member and their Jakarta facility qualifies as a Tier 3 teleport.

7.3.7 The Intelligent Island: Singapore

As mentioned in Section 7.3.6, Kakuchopurei operates several teleports in the Southeast Asia region, including Singapore.

There is also the main Singapore communications services provider known as Singapore Telecommunications Limited (Singtel) [8], which owns and operates the teleport known as the Bukit Timah Satellite Earth Station. Singtel is a WTA member. Other Singapore teleports (all owned and operated by Singtel) include:

- Seletar Earth Station (another Tier 4 WTA certified teleport);
- Sentosa Satellite Earth Station, Sentosa Island;
- Tampines Telepark, Singapore.

7.3.8 South Korea Teleports

KT Corporation [9] is the major telecom operator in South Korea and the company owns and operates two teleports: KT Yongin (Boeun) Teleport, and KT Geumsan (Kumsan) Teleport.

Yongin is located in the Daejeon area, while Geumsan is installed in the Seoul metro area. Due to its unique function, the first installation to be considered is the Busan Maritime Business Center. This center provides the following services:

- Inmarsat service and also MVSAT (Unlimited Maritime VSAT) service contributes to efficient and stable ship operation and improvement in the welfare of sailors and crew members.
- The usefulness of the maritime satellite communication service of global Ku bandwidth provided by using KOREASAT Satellite and Global Satellite Network is introducing a new maritime culture.
- The center is organized with experts in maritime and ship for marketing, sales, and consulting. Also, the center works jointly with Kumsan and Yongin Satellite Control Centers for various field services such as the installation of communication equipment within ships, training, and maintenance.

The following provides a brief history of the Yongin Satellite Control Center development stages.

In November 1994, construction was completed for Yongin/Daejeon Control Center and Control facility. The Yongin Satellite Control Center was opened, and, in August 1995, Koreasat 1 was launched, followed by nine further launches spread over the period 1996 to 2017 when Koreasats 7 and 5A were successfully launched.

Following these launches:

- In 2015, in cooperation with the Electronics and Telecommunications Research Institute (ETRI), the Satellite Control System was developed.
- In 2016, the Implementation of Global Network monitoring site was completed.

All these developments place Korea (especially the KT Corporation) in a strong position for further business advancement as New Space becomes a profitable reality.

7.3.9 Teleports in the United States and Canada

Unsurprisingly, the United States has the largest proportion (per capita) of teleports in the world. Some interesting specific North American teleports are described in the following sections.

7.3.9.1 Teleports Previously Owned and Operated by COMSAT

Until recently, a U.S. company previously known as COMSAT owned and operated two teleports in the United States: one in Southbury, Connecticut, and the second in Los Angeles, California (both Tier 4). Then, in March 2023, a U.K.-headquartered company known as Goonhilly Earth Station Ltd (GES) [10], based in Cornwall, United Kingdom, acquired both sites. GES stated that this acquisition enables them to increase their presence in the international satellite communication market, taking the company closer to achieving 24/7 global communication coverage.

7.3.9.2 West Cedar Hill Teleport, Texas

This is an example of a notably large teleport, but then most things are surely larger than life in Texas. By 2023, there were at least 23 dishes of various diameters on this site.

Westar Satellite Services [11], a Texas-based satellite communications company, operates this teleport near Dallas. It is used for data and video transmissions, transmitted using other satellite carriers' networks and through the company's ground-based networks connecting Texas' major cities. It is also close to some large radio and TV broadcast towers.

7.3.9.3 Brewster (Washington) and Vernon Valley (New Jersey) Teleports

Both teleports are WTA Tier 4-qualified and are owned and operated by the American company U.S. Electrodynamics, Inc. (USEI) [12], itself headquartered in the city of Brewster, Washington. USEI's teleports are located in Brewster (4 miles north of the city center) and Vernon Valley.

By 2023, there were at least 30 dishes of various diameters on the Brewster site. Most are Ku-band, but 7 are C-band. The Vernon Valley site is smaller, with 13 Ku-band dishes and 6 C-band dishes.

7.3.9.4 Canada

Most of Canada's teleports, located west to east across the country (from Winnipeg to Ottawa), are owned and operated by Telesat, headquartered in Ottawa [13]. Telesat also owns and operates the Anik and Telstar series of satellites.

7.3.10 U.K.-Based Teleports and Related Operations

As mentioned in Section 7.3.9.1, GES (Goonhilly) [10] has emerged to become an aggressive U.K. teleport player active internationally. Other important U.K.-based companies include Arquiva and Talia.

7.3.10.1 Arquiva

Arquiva [14] is headquartered in Gerrards Cross (a few miles west of London) and operates a total of 4 teleports variously located around Southeast England.

One of the largest providers of teleport facilities in Europe, Arquiva operates more than 150 antennas from their four main teleports. This provides access to a large number of satellites capable of meeting all Arquiva's customers' transmission needs reliably, efficiently, and securely.

Colocated at the company's headquarters, there is the Chalfont Grove Teleport (WTA Tier 4). The other Arquiva U.K. teleports are:

- Bedford Teleport (Bedford, due north of London);
- Morn Hill Teleport (Winchester, well to the south of London);
- Martlesham Teleport (Ipswich, northeast of London);
- Crawley Court Teleport (also like Morn Hill in Winchester).

This company maintains a significant presence in Ireland, mainland Europe, and the United States. Customers include major broadcasters such as the BBC, ITV, BSkyB, and the independent radio groups, major telco providers including the United Kingdom's five mobile network operators, plus the emergency services.

7.3.10.2 *Talia Teleports*

Although Talia [15] is headquartered in London, the only teleport owned and operated by this company seems to be located in Raisting (WTA Tier 3, Germany).

7.3.11 Italy

7.3.11.1 *Telespazio (in Italy)*

Substantial information about this company is provided in Section 7.3.3. In Italy, four major teleports are owned and operated by Telespazio [3]: "Piero Fanti" Space Centre in Fucino (L'Aquila) and the Matera, Lario, and Scanzano spaceports.

The Fucina Space Centre, the largest and most extensive in Italy, had been operating for over 50 years when, in 2018, it received the prestigious full certification assigned by the WTA (i.e., Tier 4).

Telespazio's "Piero Fanti" Space Centre in Fucino (L'Aquila) has been active since 1963 and with its 170 antennas and 370,000 square meters, it is now recognized as the first and most important teleport in the world for civilian use.

The Fucino Space Centre performs in-orbit satellite control, space mission management and telecommunications, television, and multimedia services. Operational logistics and field services are active in support to the services provided. The teleport employs 250 workers including engineers, specialist technicians, and operational staff. Fucino hosts the Control Centre and the Mission Centre of the COSMO-SkyMed Earth observation satellite constellation and one of the two Control Centres that manage Galileo, the European satellite positioning and navigation system. The Italian Galileo Control Centre (GCC-I) is an infrastructure of 6,000 square meters that ensures processing and distribution of the navigation signal to satellites and continuous control of the quality of service delivered to end users. This center also manages the Galileo Data Dissemination Network (GDDN), which includes about 50 ground stations.

7.3.11.2 *GlobalTT*

GlobalTT [16] is a European company headquartered in Brussels, Belgium. It provides relatively simple and also complex internet satellite solutions covering the entire world, but with a main focus on Africa. The company began in 1994, developing new solutions and performing on-site work to provide a secure service to its clients. As a private satellite Earth station, GlobalTT operate 100% of their business on a same site comprising antennas, fiber links, broadband platforms, security, support team, sales team, logistics, and both software and hardware development. The company also monitors quality control (24/7), management, and stock, all on the same site.

As an operator, GlobalTT uses and manages the bandwidth of 10 satellites to cover clients requiring the Ku-band/C-band over Africa, the Ku-band over the Middle East, and the Ka-band over Europe. It is well known that businesses, nongovernmental organizations (NGOs), embassies, and projects in many regions of Padua struggle with internet access. GlobalTT/IPSEOS is a seasoned satellite operator with over 30 years of experience in Africa. It provides reliable solutions that help to overcome these challenges with reliable connectivity, protection against cyber-attacks, and optimized management of internet applications.

7.3.12 Other Teleports: Mainly Located in Regions Not Considered Above

The teleports considered in the above sections comprise detailed examples of some specific installations. Further teleports are now identified (Table 7.1) located a various countries around the world.

The main criterion determining the selection in Table 7.1 is the relatively high-quality service offered in each instance.

7.3.13 Africa

Although generally considered a region relatively underserved in terms of telecommunications, there are at least 30 teleports mainly dotted around the coastal areas of this huge continent. Until recently, the main issue was the relatively poor quality of the services from these teleports. However, in May 2023, a new gateway was installed by Avanti Communications in Dakar, Senegal (West Africa) and, when fully operational, this promises:

1. To greatly improve the quality of communications;
2. To form an important example for other parts of Africa to follow.

Table 7.1
Global Distribution of Selected Other Teleports

Location	Operator
Europe	
Vienna, Austria	A1 Telekom
Bonn and Koln, Germany	AXESS
Mooslurg a.d. Isar, Germany	Horizon Teleport
Baku, Azerbaijan	AzerCosmos
London, United Kingdom, and Oslo, Norway	Telenor
Ljubljana, Slovenia	STN
Makirios, Cyprus	Cytaglobal
Santander, Spain	Santander Teleport
Sofia, Bulgaria	Vivacom
Central and South America	
Mexico City	AXESS
Bogota, Columbia	AXESS
Australasia*	Optus (Australia headquartered)
India†	PlanetCast
Jakarta, Indonesia	Primacom Interbunae

Notes:
*Australia, New Guinea, and New Zealand.
† Cochin, Mumbai, and New Delhi.

In the fall of 2023 this installation mainly comprised the 2023 American-built Ka-band antenna that has the following physical characteristics:

- Diameter: 9.2m;
- Height above ground: 14m;
- Weight: 17,000 kg.

This sophisticated antenna took more than a year to manufacture and, when fully up and running, will provide state-of-the-art backhaul and large-scale connectivity for commercial telecoms and government agencies.

Avanti's local gateway partner is the Senegal CSP known as Free, and this company will host and support the operations of the new gateway from its Tier 3 data center facility in Diamniadio. It seems fair to state that this HYLAS 4 gateway is critical to Avanti's plans to connect underserved communities in Senegal. This project will also provide a strong stimulus towards the implementation of many further such sites across Africa [17]. For example, Avanti has connected more than 1,000 villages and schools across the continent and has ambitious plans to connect a further 10,000 sites at least through 2028. HYLAS-4 is an example of a completed (Avanti) gateway.

7.4 LIKELY FUTURE SCENARIOS FOR TELEPORTS

In 2023, practically every teleport in the world was using microwave or millimeter-wave uplinks and downlinks. Bands included the C-band (microwave radio) or Ku-band, both of which are microwave. More recently, mmWave links have been implemented, notably, the Ka-band, and the Senegal gateway antenna described in Section 7.3.13 is a good example. The possible use of higher mmWave bands operating at many tens of gigahertz has been considered, but the lack of sufficient transmitter power even for the uplink is a substantial issue.

So the fundamental question is being asked and indeed addressed: What about FSO links? Such a link has already been demonstrated in Western Australia and a picture of the optical Earth station (including the beam) is shown in Figure 7.5.

More recent information [18] has indicated the progress being made in this sphere.

How a particular teleport develops, and even what form of teleport is envisioned for a specific region, depends critically upon the pattern of user needs and the local economic circumstances at the time of conception and during the early-build phase. As described in Section 7.1, it is possible to define four basic forms: the telecom-port, the real-estate teleport, the intelligent city, and, finally, the most advanced concept of all, the virtual teleport. Well into the third millennium, it is clear that the world is continuing to see examples of each of these basic forms being established in various locations.

Further, it is important to appreciate that existing teleport projects will continue evolving so that projects that began as telecomports or real-estate teleports will become essentially intelligent cities.

Virtual teleports are different in that they virtually integrate existing advanced telecommunications infrastructures, usually with some added value from introduced subsystems and services. Cities ripe for this form of development are always those that already have

Figure 7.5 The TeraNet-1 optical ground station at the University of Western Australia [1].

a sophisticated information infrastructure but not formally a teleport. A range of potentially suitable cities is provided in the first edition of this book, but in the 23 years that have elapsed since that time, most indeed developed to include teleports. These success stories include Austin, Texas; Buenos Aires, Argentina; Cape Town, South Africa; Delhi, India; Denver, Colorado; Guangzhau, China, PRC; Hong Kong, China, PRC; Los Angeles, California; Madrid, Spain; San Diego, California; Singapore; Sydney, Australia; Stockholm, Sweden; and Berlin, Germany. Examples of cities expected to acquire teleports in the near future include Birmingham, United Kingdom; Dresden, Germany; Dubai, United Arab Emirates; Glasgow, United Kingdom; and Leeds, United Kingdom. Tel Aviv, Israel's main commercial center, has a teleport, but this requires updating and may indeed be held back until the country's war with Hamas is over.

Again, concerning the next two or three decades, many cities in the G8 regions will demand teleports. Some may be more or less conventional (i.e., microwave/millimeter-wave) connected with the familiar parks exhibiting several dish reflector antennas and FSO links will become increasingly installed.

In second-world and third-world nations, the situation is naturally different and more complex. India, for example, has a strong legacy in satellite communications and advanced industry, and yet many of its city centers still need substantial redevelopment. Major cities such as Delhi and Calcutta now boast teleports or real-estate teleport projects.

In regions such as Central and South America, apart from Brazil where the Rio de Janeiro Teleport is developing, there is almost no history of advanced industry but companies such as Telespazio have already installed teleports (see Section 7.3.3).

Global operators are likely to emerge to offer teleport project development and management at any level suited to the local situation and budget. This has already started by the WTA that has considerable consultancy experience in this field.

The concept of a teleport enterprise corporation (Global TEC) may become relevant here. Such an organization would provide a highly flexible service such that it could efficiently and dynamically manage practically any form of teleport project in the world. An impression of this concept is given in Figure 7.6.

In this speculated schematic four operating projects are indicated, one in each of four countries: India, South America, Germany, and the United States. It must be emphasized that this is purely speculative and that, as far as is known, no such projects actually exist at the time of this writing. In Figure 7.6, the three nebulous bubbles are intended to indicate potential projects that have not so far reached the signed-up stage. All operating and potential projects involve two-way highly interactive dialogue, as shown by the arrows. This dialogue would be efficiently run via telecoms services companies as far as possible with occasional physical visits, although only where absolutely necessary.

Most likely, the main focus should be on common language connections in which those nations that share a largely common language may have preferential channels between teleports in various countries. Australia, New Zealand, North America, and the United Kingdom would feature strongly in this respect with English being

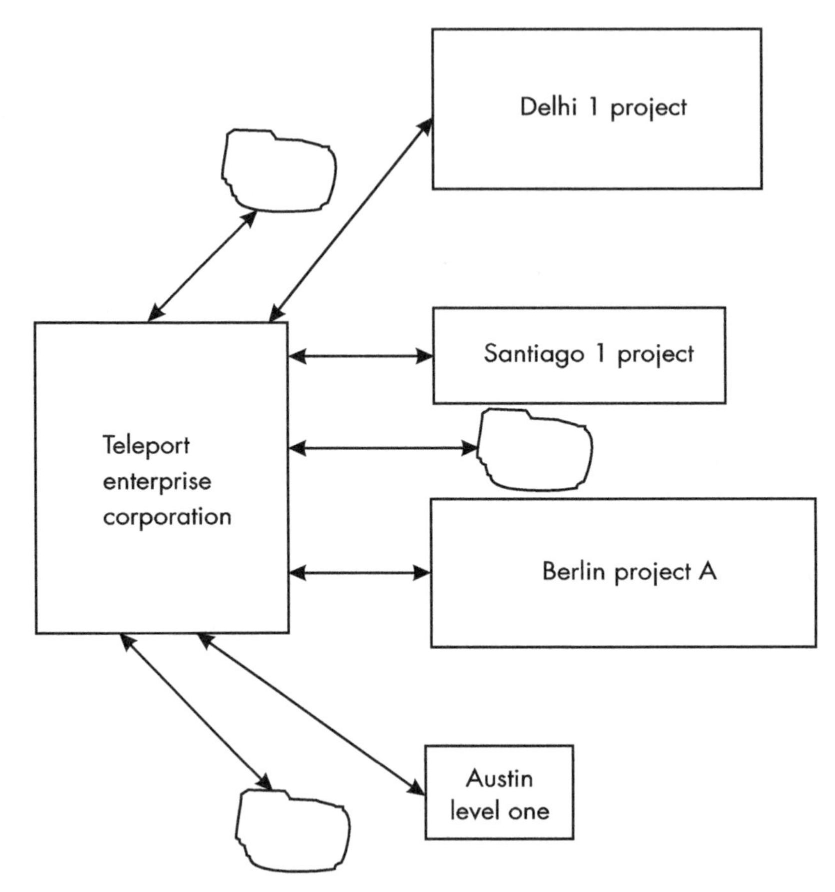

Figure 7.6 The teleport enterprise corporation concept. *From:* [20].

the working language. Chinese would also be increasingly important, mainly covering the vast expanse of China, and Spanish would be an added language of great significance throughout Central and South America, as well as Spain itself. Portuguese would also probably be important, for Portugal but also for Portuguese-speaking Brazil.

As projects develop to sufficient levels of maturity so the teleports themselves would be used to increasing extents as on-ramps for superhighway communications. As more sophisticated SATCOM systems come online such as, particularly, LEO SATCOM systems, a gradual trend away from teleports as such may well gain traction.

7.5 SOME TECHNOLOGICAL ASPECTS

7.5.1 Antennas

For the past many decades, the main types of antennas used in teleports (also VSATs and many types of gateways) have involved parabolic reflectors. This fact is amply indicated when observing all the teleports described above with the important exception of optical ground stations. Increasingly, however, flat-panel antennas are beginning to enter service for RF links rather than the more familiar parabolic dish reflectors.

7.5.2 Software-Defined Radios

Digital technology accompanied by the application of appropriate software are both increasing in importance for all applications, including teleports. SDRs, described in Chapter 1, are a significant example of this trend. Reflector antennas, flat-panel type antennas. and SDRs have all been described in some detail by Edwards [19].

References

[1] Varrall, G., *5G and Satellite RF and Optical Integration*, Norwood, MA: Artech House, 2023.

[2] www.worldteleport.org.

[3] www.telespazio.com.

[4] www.a-cosmos.co.jp.

[5] www.aruji.com.

[6] https://www.kakuchopurei.com/2022/08/genshin-impact-teleport-way-point-located-in-malaysia-singapore-indonesia-philippines/.

[7] https://primacom.com/.

[8] www.singtel.com.

[9] https://corp.kt.com/eng/html/intro/main.html.

[10] https://www.goonhilly.org/.

[11] http://www.allmobilevideo.com/satellite-services/.

[12] https://www.usei-teleport.com/.

[13] www.telesat.com.

[14] www.arquiva.com.

[15] http://www.talia.net.

[16] https://www.globaltt.com/.

[17] https://www.connectingafrica.com/author.asp?doc_id=785083&itc=newsletter_connectingafricainsights&utm_source=newsletter_connectingafrica_connectingafricainsights&utm_medium=email&utm_campaign=05262023.

[18] https://www.laserfocusworld.com/lasers-sources/article/14293795/laser-space-comms-was-missing-link-for-space-science-instruments.

[19] Edwards, T., *Technologies for RF Systems*, Norwood, MA: Artech House, 2018.

[20] Edwards, T., *Gigahertz and Terahertz Technologies for Broadband Communications,* 1st ed., Norwood, MA: Artech House, 2000.

8

TERRESTRIAL AND STRATOSPHERIC BROADBAND WIRELESS

8.1 BROADBAND WIRELESS WITHOUT SATELLITES OR FIBER

Most of the high-profile advances in broadband and high bit rate communications, increasingly for internet (5G) on-ramps, are the province of new and planned satellite constellations and DWDM fiber systems. The latter were described in Chapter 3 and broadband satellite systems are discussed in Chapter 9.

However, neither satellite nor fiber represents the only possibilities for the delivery of broadband wireless services—far from it. It is well known that microwave or millimeter-wave signals will travel through the atmosphere, and this is how it all began. Usually, the term *microwave* tends to mean to most people one of two possibilities: either line-of-sight terrestrial communications or else something one cooked with very efficiently.

Before fiber optics gained the undisputed lead in terrestrial communications systems, line-of-sight microwave links with their dish-festooned towers were and still are clearly visible across most industrial countries. Prior to the 1980s, most links were analog, using FM and FDM, but, by the 1990s, digital transmission was being implemented quite generally. Most of these types of systems, now almost entirely digital, remain in operation and also the installation of new microwave or (increasingly) millimeter-wave links continues unabated.

This approach is vital in regions where it would be completely uneconomic to install fiber-optic cables due to either low-population density or rigorous terrain, or both these factors. These types of links often involve very broad overall bandwidths, to accommodate large numbers of channels, but the individual basic services are often of a relatively narrowband nature (e.g., voice or text).

Why is it that neither satellite systems nor fiber-optic transmission can be expected to deliver all the broadband services required, well into the third millennium? This is an especially interesting question because, ultimately, most communications traffic and, indeed, most information processing will almost certainly be by photonic means, with wireless taking up the mobile needs locally. But, this scenario is still several years away, and we deal with some of these developments and issues in Chapter 10.

Wireless provides an excellent solution for the last mile (e.g., local access). The key question is, will satellite systems not always prove to be the decided winner for customers living and working in regions of low-population density or difficult terrain? This may be the case when humanity approaches the mid-twenty-first century (see Chapters 7 and 9). Before the New Space era matures, terrestrial wireless continues to provide excellent solutions for direct to the customer connections as well as providing backhaul and similar solutions in the radio access networks (RAN) for 5G and the coming 6G era.

Another feature applicable to both fiber optics and satellite systems is the time to deployment. Entirely new fiber-optic cabling projects take years to plan and install. As mentioned in Chapter 3, where existing fiber cabling installations already have narrowband multimode fibers, these can be fairly readily upgraded for the broadband future at dramatically reduced costs to the provider. Eventually, as existing cables age, this will become a further tremendous advantage for cabled networks.

Regarding satellite-based systems, a factor to be considered is the considerable downtime and cost of any maintenance. It is generally far too expensive to think in terms of astronauts servicing spacecraft, but specialized robotic spacecraft can perform some of the anticipated faults when these occur on satellites. It is important to note that broadband satellite will largely be suitable for mobile users and this is a vital advantage.

Even given the strong growth in mobile users with their demands for on-the-move communications, there will continue to be large numbers of customers who need broadband telecommunications, notably for internet access, from their homes and workplaces. This requirement is driven onward by the fact that these homes and workplaces are increasingly identical locations.

Observing the above challenges for both fiber cabling and satellite communications, several substantial industry players have, so to speak, brought things closer down toward the ground, and, by the end of the twentieth century, strong growth was well established in terrestrial multipoint/multimedia systems. HAPS (broadband wireless with stratospheric platforms) are also described later in this chapter.

8.2 FREQUENCY BANDS

It is vital to understand the distinctions between microwave, millimeter-wave, and sub-millimeter-wave. Strictly speaking, as far as 5G is concerned, the distinctions follow the following frequency ranges (all in gigahertz):

- Microwave: up to 23.9 GHz;
- Millimeter-wave: 24 to 140 GHz;
- Sub-millimeter-wave: above 140 GHz.

In practice, the lower 5G bands (FR1 and FR2) stop at 6 GHz.

Notice that these practical ranges differ from the classical frequency bands, which follow the end frequencies: 3 GHz, 30 GHz, 300 GHz, and so on.

Microwave frequency bands are defined as follows:

L-band: 1.12 to 2.25 GHz;

S-band: 2.26 to 3.94 GHz;

C-band: 3.95 to 8.19 GHz;

X-band: 8.2 to 12.4 GHz;

Ku-band: 12.41 to 18 GHz.

Details concerning the millimeter-wave frequency bands are given in Tables 8.1 and 8.2.

8.3 THE NETWORKS

Cellular (mobile) networks generally comprise the following vital elements:

- Personal mobile handsets (cell phones);
- The RAN as such, connecting core sites using backhaul, fronthaul, and midhaul (xHaul) using microwave or increasingly mmWave links in mesh configurations;
- The core sites themselves;
- Rapidly addressable memory installations (especially the cloud).

In many instances, the final connections to the subscribers' premises is provided by fiber-optic cabling, but 60-GHz fixed wireless access (FWA) is increasingly important for last-mile connectivity, requiring mmWave radio. Each site will have its own internal communications access and will be equipped with further antennas enabling two-way broadband communications between antenna towers. All these subsystems comprise backhaul, fronthaul, and midhaul (collectively, xHaul) using microwave, mmWave, or fiber-optic links.

Table 8.1
K-Band to E-Band Millimeter-Wave Bands (all GHz)

Letter	K-Band	Ka-Band	Q-Band	U-Band	V-Band	E-Band
Frequency	18–26.5	26.5–40	33–50	40–60	50–75	70–80

Note: In practice, millimeter-wave tends to start at 24 GHz.

Table 8.2
W-Band to D-Band Millimeter-Wave Bands (all GHz)

	W-Band	F-Band	D-Band
Frequency	75–114	90–110	110–170

This chapter focuses on mmWave links for these applications mainly using the licensed E-band (i.e., 70 to 80 GHz).

Mesh-configured RANs are increasingly being adopted mainly because they are optimized multipoint-to-multipoint, which substantially decreases the number of radios required.

Efficient access to the cloud is always of critical importance and the Open RAN (O-RAN) is now highly significant. There is an O-RAN Alliance comprising several company members and O-RAN is designed to accommodate a variety of suppliers' types of radios. Edge computing, leading to lower latency, represents an important aspect. During 2023, Agilent Technologies began testing O-RAN networks in Europe.

FWA uses mmWave technology directly installed in people's homes, with the other end of the radio link connected to an appropriate RAN cell in the 5G core network. In the past, most FWA installations used the K/Ka-bands, but the application of the unlicensed V-band (60-GHz center frequency) is rapidly gaining acceptance. This is a particularly interesting question because V-band technology had already become unattractive for xHaul.

In this industry, the term "xHaul" refers to the connections between elements of the cellular networks. xHaul includes backhaul, fronthaul, and midhaul. Basically, the technology for xHaul has become focused on the E-band where the bandwidth is considerably wider than V-band.

8.3.1 Advanced Mesh Networks

First, it is necessary to describe mesh networks as such, beginning with a basic and quite general definition. A LAN is defined as a mesh network when the infrastructure nodes (computers, bridges, switches, radios) connect directly, dynamically, and nonhierarchically to as many other nodes as possible and cooperate with one another to efficiently route data to and from clients. It seems likely that advanced (wireless) mesh networks will steadily become the de facto standard from around the mid-2020s onwards.

8.3.2 Radios in xHaul Links

The radio environment associated with xHaul is increasingly characterized by the following aspects:

- The increasing adoption of E-band;
- The increasing adoption of mesh networks;
- Increasingly high densities of mmWave radios.

Every radio (i.e., every node) must be able to interact with every other node and this places tight specifications on the technology. An example of a fairly typical mmWave radio is shown in Figure 8.1.

The full-duplex radio shown in Figure 8.1 is the Siklu EtherHaul 8010FX (EH-8010FX), which provides 10-Gbps transmission speed across the 70/80-GHz E-band spectrum.

8.3.3 Radios in Fixed Wireless Access Links

The radio environment associated with FWA is characterized by the increasing adoption of V-band (centered on 60 GHz). Another major feature associated with FWA is the fact that this subsystem is directly connected to the final end-user, the domestic/business customer (or an organized group of customers). This means that the FWA radio

Figure 8.1 A typical commercial mmWave radio. (Courtesy of Siklu and thanks to Tim Herbert, who provided this photo.)

does not have to connect to any other nodes and the radio can therefore be offered at a lower price compared with xHaul.

8.3.4 Radio Propagation in the Earth's Lower Atmosphere (the Troposphere)

This is where most of the design and implementation issues tend to reside when contemplating the design of terrestrial microwave or mmWave RF (radio) links. Clearly the RF beam has to travel through Earth's lower atmosphere point-to-point or point-to-multipoint from antenna to antenna located at the specific Earthbound points. Because this lower atmosphere contains both free oxygen and water vapor (both of which affect radio propagation) transmission losses in decibels per kilometer change very dramatically as frequency changes.

Across all frequencies, DC right up though cosmic radiation, the transmission losses versus frequency is called the electromagnetic spectrum. The RF region is a relatively small part of this, although it is very important in terms of radio communications because of new 5G and also 6G. Details concerning the overall attenuation (in decibels per kilometer) are well known and are shown in Figure 8.2. These apply to sea-level conditions and with the antennas directed vertically through the atmosphere.

Over the full frequency range from 15 to 400 GHz, there are clearly five peaks in attenuation (resonances) caused by the molecular absorption of water (H_2O) and oxygen (O_2) in the atmosphere. In addition, the omnipresence of mainly nitrogen comprising the atmosphere itself is increasingly being absorbed as frequency increases. These features result in the general curve.

Through microwave frequency bands, up to about 18 GHz, the atmospheric attenuation is relatively small and increases only slowly with frequency, reaching approximately 0.02 dB/km at 18 GHz. This covers, for example, all the frequencies associated with microwave 5G (i.e., up to 6 GHz). There is a significant water vapor resonance (increased RF power absorption) at 22 GHz, and this explains the accelerating rise in attenuation reaching a maximum of 0.35 dB/km at 22 GHz. Between about 26 and 45 GHz, the attenuation is relatively flat, remaining below about 0.2 dB/km. This region includes the important Ka-band of frequencies. Above 45 GHz, the attenuation begins to rise steeply, reaching a maximum of almost 12 dB/km at 60 GHz. This is the first peak due to molecular oxygen absorption, and this

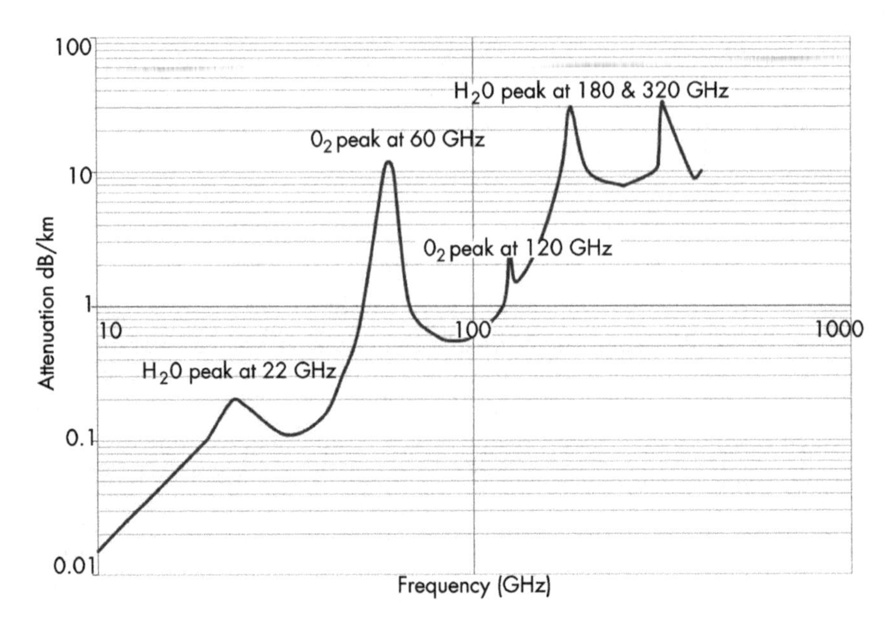

Figure 8.2 Radio attenuation (dB/km) over a wide range of frequencies in dry air. (Originally generated by Chris Gibbins at Rutherford Appleton Laboratory, United Kingdom.)

peak occurs at the top of a highly nonlinear resonance around 60 GHz. FWA installations increasingly use this band, but these cover relatively short ranges, typically a few hundred kilometers maximum, so the increased attenuation is tolerable.

As frequency is increased, the remainder of the spectrum exhibits repeated resonances, and generally frequency bands useful for RF communications tend to include the dips in attenuation. This principle applies throughout the millimeter-wave and sub-millimeter-wave (terahertz) bands, so it is an important consideration for 5G and 6G (see Chapter 10). An application program named MATLAB includes a built-in function called "gaspl" that uses ITU-R recommendation P.676 to calculate the gas attenuation applying to radio propagation [1].

Another important aspect applying to the design and implementation of terrestrial microwave or mmWave RF (radio) links concerns the antennas. Due to features such as ground reflections, it is necessary to implement dish-reflector antennas. Flat-panel (phased-array) antennas are unsuitable due to sidelobes and ground reflections.

In most instances, SDRs are implemented and, during the twenty-first century, cognitive radios may become the de facto standard. SDRs were described in Chapter 1. Cognitive radio comprises a radio system that has been programmed and dynamically configured to select the best RF channels available [2, pp. 9–10].

8.3.5 Total Cost of Ownership

Fiber optics inherently provides a huge bandwidth capability, orders of magnitude greater than possible with the best conceivable radio system. Therefore, it ought to be obvious that this technology should generally be adopted where possible. In this case, the total cost of ownership (TCO) is limited to the following:

- The cost of leasing the fiber;
- Any additional leasing or ownership costs associated with being a user of the RAN network as such.

However, in many instances, fiber is not immediately available, and this leads to the adoption of radio links. It also means the TCO is much smaller than for the fiber cable situation. In addition, the greater the distances involved, the greater the savings will be, although most radio links are distance-limited.

8.3.6 What About Standards?

In most areas of widespread human activities, it is generally vital to establish standards. All modern forms of telecommunications include this requirement. With 5G, new radio standards fall into the following basic groups:

- FR1 (C-band): Adopting the (relatively) low frequency bands, generally known as sub-6 GHz;
- FR2: Utilizing higher frequency bands, generally above 24 GHz.

There are also the following important millimeter-wave bands: V-band (60 GHz center) and E-band (70 to 80 GHz).

During the early 2020s, the use of both V-band and E-band was growing very strongly with V-band being used for FWA, while E-band represented an almost equally strongly growing option for backhaul, fronthaul, and midhaul 5G network connections. Within this book, further information is provided concerning these important developments. Chapter 1 of Geoff Varrall's recent book *5G and Satellite RF and Optical Integration* [3] provides a substantial and detailed coverage of relevant 5G standards, including extensive tables.

For 5G new radio, the needs and details concerning appropriate standards are developed and announced by organizations such as the ITU, the IEEE, and the World Radio Conference (WRC).

Toward 6G (described in Chapter 10), substantial activity is being focused on the use of subterahertz frequencies and it is widely appreciated that this is a very challenging field of technology. Also, such frequencies result in tightly focused radio beams that are incompatible with reception on advanced smartphones. As a result of these constraints, during 2023, there was renewed activity towards the mid band (approximately 7–24 GHz) for 6G primary operating frequencies. This band is unofficially termed FR3.

8.3.7 Markets for mmWave Terrestrial Radio Systems

According to most analysts during the early 2020s, the total available markets (global TAMs in U.S. billions of dollars) were in the range of $8 billion to $10 billion. By 2030, it has been forecasted that this could increase to about $13.6 billion and typical data points are shown in Figure 8.3.

The data in Figure 8.3 includes millimeter-wave radios adopting the following frequency bands:

- K/Ka-band: 24 to 30 GHz (lower-band FWA);
- V-band: 60 GHz (upper-band FWA);
- E-band: 70 to 80 GHz (xHaul).

Both FWA and xHaul markets are currently strong and exhibit substantial growth albeit at different overall and short-term rates. It seems clear that terrestrial radio links will continue to be as important as nonterrestrial networks (NTNs). By 2023, it also became clear that interconnections between NTNs and terrestrial communications were already becoming important, and this trend will likely continue through the twenty-first century.

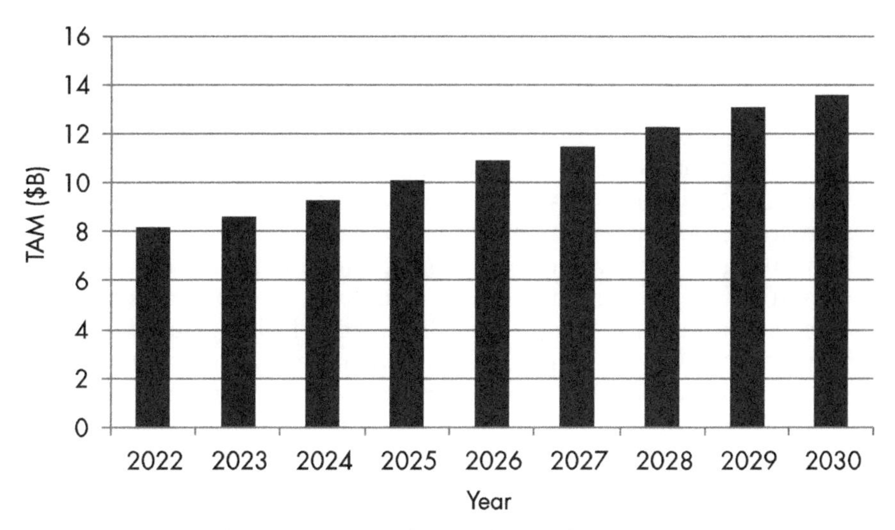

Figure 8.3 Global TAMs (billions of U.S. dollars) for mmWave terrestrial radios (data provided by the author).

8.4 HIGH-ALTITUDE PLATFORM SERVICES

8.4.1 Gaining Early Traction in the 1990s

Toward the end of the twentieth century, it was appreciated that neither terrestrial nor satellite-based systems (or indeed combinations of these) were the only technologies for broadband communications. Terrestrial networks are, by definition, installed in the Earth's troposphere (i.e., the region of densest air in which the great majority of us live and communications technologies appropriate to this region are the subject of most of the above material). Contrastingly, communications satellite constellations orbit at altitudes ranging from several hundred kilometers (LEOs) out to several tens of thousands of kilometers (GEOs). These are all described in Chapters 4, 7, and 9.

Between the extremes of terrestrial and space, there is the outer region of Earth's atmosphere known as the stratosphere. Although the air density here is relatively small, balloons and also very lightweight aircraft can be stationed in this region.

Some details concerning developments in the 1990s were provided in the first edition of this book and, because it is highly relevant selected material from that book, is presented here in annotated form. In particular, the tense is typically converted to past tense because this emphasizes context:

Various players are actively interested in deploying stratospheric balloons as platforms for broadband services. Probably the best known was Sky Station International who intended to fly their first commercial balloon for this purpose in year 2000. Sky Station's ultimate plan was to have a fleet of about 250 balloons in service, hovering at 21-km altitude above many major cities, at some time before 2005. Each balloon would carry a payload up to a maximum of around 1,000 kg and power was obtained from a combination of fuel cells and solar arrays. The 21-km altitude can readily be imagined, being just over twice the typical altitude (marginally below 10 km) at which we are all familiar when flying in commercial jets. By 2005 these fleets of balloons and other craft would probably have become familiar sights for aircrews, their passengers, and indeed many people on the ground.

A substantial industrial consortium comprising Aerospatiale, Dornier Satellitensysteme, and the Jet Propulsion Laboratory (owned by NASA) was in place to pursue this program. Technological problems that were yet to be solved included the difficulties of operating in the extremely cold environment—it is not often appreciated that the upper stratosphere is actually colder than outer space because of conduction and convection. There are also the not inconsiderable difficulties involved in maintenance and the replenishment of consumables. Another airship-based network was proposed by the Japanese Ministry of Posts and Telecommunications, and Turin Polytechnic (Italy) was cooperating with the Italian Space Agency (ISA) to develop and construct a long-endurance aerostat powered by fuel cells for night operation and solar energy by day. At 100 kg the payload capability was only 10% of Sky Station's and the available electrical power was also relatively low. But this Turin/ISA project could still have been very much a winner given the anticipated advances in practically all electronic hardware, including microwave and millimeter-wave solid state device (MMIC) capabilities and their ever-decreasing supply power requirements.

Also, the Turin/ISA program was expected to only involve a $3 million investment and to cost $345 per hour to run. For a start, the $3 million was at least an order of magnitude lower than any planned terrestrial system at the time and around two orders of magnitude below the investment levels associated with most broadband satellite projects.

> Yet another project, known as RotoStar, was being developed at Tel-Aviv University in Israel. Like both Sky Station and Turin/ISA, RotoStar was designed to hover at the 21-km altitude. Unlike any of the other schemes publicized to date, the platform vehicle for this project would remain on station for periods between four days and six months. A critical distinction with RotoStar was the fact that it was designed as an unmanned rotating-wing craft rather than strictly a balloon.

In the 1990s, fleets of high-flying aircraft (high altitude long operation (HALO)) were also serious considered with each aircraft carrying broadband transponders.

> Since 1996, the Raytheon Company had been working with another corporation known as Angel Technologies to demonstrate the feasibility of providing broadband wireless services using HALO aircraft. By late summer 1998, the two players entered into a formal teaming agreement for this program and demonstrated the basic feasibility of the concept. Angel Technologies' HALO aircraft design, called HALO-Proteus, could carry a large antenna high above the center of a city to deliver very high-speed communications services. Augmenting terrestrial microwave towers and orbiting satellites, the HALO aircraft would have flown fixed-pattern trajectories in the stratosphere in order to provide metropolitan wireless services at a lower cost, with increased flexibility and improved quality of service, than either satellite or fiber. A variety of fixed and mobile wireless services were offered including voice, data, and video.
>
> In August 1998, a test flight was conducted using ground and airborne electronics developed by Raytheon in which the companies jointly demonstrated a record-breaking 52-Mbps wireless link between the ground and Angels' HALO Network testbed aircraft in flight. Under the teaming agreement, Raytheon would have responsibility as the prime electronics systems integrator for both the airborne and ground segments of the HALO network. This company would leverage its extensive capabilities in complex integrated RF systems and digital technologies to architect the HALO network. By late 1998, Angel Technologies announced an agreement with Wyman-Gordon Company for the certification and production of 100 HALO-Proteus Aircraft, and in summer 1999 this aircraft made its debut at the high-profile Paris Air Show. The platforms

> comprised relays of HALO-Proteus aircraft circling for periods up to 16 hours at 16-km altitude, providing footprints on the ground (cones of commerce was the term used here) that could have diameters of up to 120 km. Twenty-four-hour broadband communications availability was provided. The aircraft intercommunicated between HALO gateways, high-capacity business premises, and domestic consumer premises. There was a network operations center and also connections out to remote metropolitan centers.
>
> For individual subscribers this type of system could have provided 1.5-Mbps internet connections at a price tag of around $40 per month early in the twenty-first century. In the case of dedicated commercial customers, connect speeds of up to 54 Mbps were promised and total network capacity would have exceeded 16 Gbps.

Today, a connect speed of merely 54 Mbps would be regarded as very modest and we would look for more like 540 Mbps. Fast forward more than two decades, and this HALO approach had moved on strongly, as we shall see next.

8.4.2 High-Altitude Platform Services or HAPS Revisited (5G High-Altitude Platform Services)

During the first quarter of the twenty-first century, the prospects for stratospheric broadband advanced strongly, while following some markedly different standards and technologies comparing the late-twentieth-century HALO described above. In particular, 5G is a major driver. Geoff Varrall devoted an entire chapter of his book to the subject, and the reader is highly recommended to check this out completely [3]. Because Chapter 9 of Varrall's book covers so much up-to-date material, we are providing relatively limited information here.

Like HALO, the new HAPS have the following characteristics:

- They can comprise either balloons or lightweight high-flying aircraft.
- They operate at altitudes between 18 and 50 km (mostly around 18–25 km).
- Microwave or mmWave links are most often chosen (see Section 8.2).

- For mmWave links particularly, atmospheric losses are significant (see Section 8.3.4).
- Free-space optical links are also feasible.

Although the altitude of a specific HAPS platform may be 20 km vertically above sea level, an interacting Earth station (fixed or mobile) could typically be something like 70 km away. In this instance, the total path loss would be at least 33 dB more than a 1-km terrestrial line-of-sight link. This feature means that the receiver on the HAPS needs to have at least 30 dB more gain than a terrestrial receiver. Obviously, this receiver begins with the antenna and at microwave frequencies this leads to a 1-m diameter antenna assuming a dish-reflector design.

The fact of their stratospheric location combined with the Earth's curvature means there are limitations regarding inter-HAPS link design as indicated in Figure 8.4.

8.4.3 AALTO High-Altitude Platform Services

For several years into the twenty-first century, the Airbus company, located in Farnborough, United Kingdom [4], had been active regarding HAPS and in 2023 the company introduced the new business name: AALTO HAPS. The company regards itself as "the world leader in stratospheric technology," and it is the developer of the

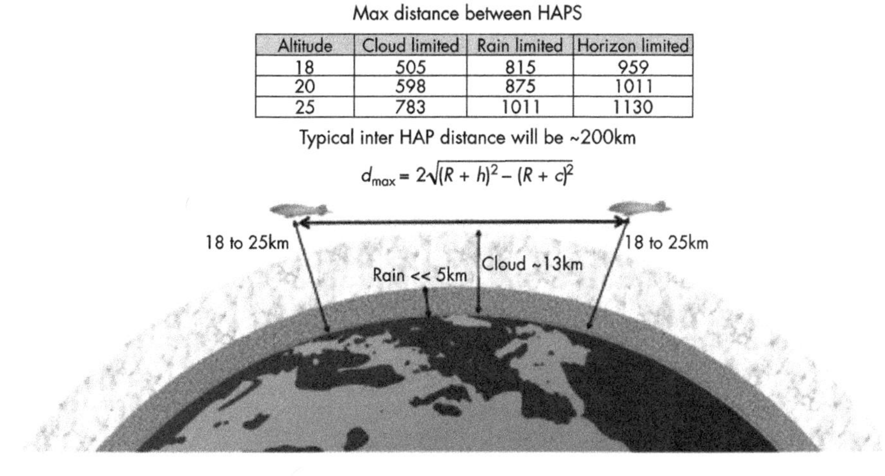

Max distance between HAPS

Altitude	Cloud limited	Rain limited	Horizon limited
18	505	815	959
20	598	875	1011
25	783	1011	1130

Figure 8.4 Distance limitations for inter-HAPS links. (*From:* [3]. Reprinted with permission.)

zero-emissions *Zephyr* aircraft. These aircraft exhibit the following characteristics:

- They are solar-powered "net-zero" and therefore very environmentally friendly.
- They fly at just over 21-km altitude.
- They fly continuously for periods of several months.
- They provide a ground coverage of 7,500 square km, which is the equivalent of up to 250 cell-phone network towers on the ground.
- They cover areas including lower population densities or difficult terrain.

AALTO also claims that their services and solutions offer multiple technological and economic advantages over legacy terrestrial and nonterrestrial solutions. It all means mobile network operators (MNOs) can expand the edge of their networks to rural and remote areas, improve their quality of service, increase their subscriber base, and reduce churn, at the same time generating healthy profits.

It will be very interesting to see how this all develops, not only AALTO but also competitors.

8.4.4 Comparisons with Terrestrial Networks and SATCOM

In the first edition of this book (pp. 190–191), comparisons with terrestrial networks and SATCOM were presented. We consider that these comparison points are still valid and important in this new age of HAPS and they are therefore repeated here.

Compared with terrestrial networks, HAPS offers the following advantages:

- Rapid deployment;
- Ubiquitous coverage from the first day of operation;
- Avoidance of local zoning restrictions relating to tower construction;
- Line-of-sight communications with almost all rooftops in each city;
- A flexible and readily upgradeable network.

Also, compared with satellite systems, HAPS has the following advantages:

- Between 20 and 2,000 times closer proximity to the users;
- An order of magnitude greater available electrical power on-board;
- Substantial capacity that can be allocated directly over densely populated regions;
- Again, HAPS comprises a flexible and readily upgradeable network;
- With HAPS, financing is on a one-market-at-a-time basis.

As a direct result of the relatively close proximity to users cited above, the round-trip time delay with HAPS is similar to long-range terrestrial networks, being of the order of tens of microseconds, unlike the many milliseconds of latency for LEOs. This is clearly a highly significant advantage.

Broadband satellite-based communications are very important, and this leads us into Chapter 9 where the details applying to several such systems are described.

References

[1] https://uk.mathworks.com/help/phased/ref/gaspl.html.

[2] Edwards, T., *Technologies for RF Systems*, Norwood, MA: Artech House, 2018.

[3] Varrall, G., *5G and Satellite RF and Optical Integration*, Chapter 9, Norwood, MA: Artech House, 2023.

[4] https://www.aaltohaps.com/.

9

SATCOM AND NEW SPACE

9.1 SOME BACKGROUND ON COMMERCIAL SATELLITE COMMUNICATIONS

9.1.1 Relevant History

Through the mid-twentieth century, most national and international communications were truly wired. Armored copper cables containing up to hundreds of copper wires covered the industrialized countries and crossed oceans beneath the sea. These cables enabled several simultaneous telephone conversations to take place via electromechanically switched exchanges (telephone switching centers), and, until well after World War II, such cables served communities satisfactorily. Well into the twenty-first century, although fiber-optic cables now dominate, copper cables are still often used and new metallic cables are frequently installed. The exception to cables was short-wave high-frequency (HF) radio, but this was unreliable and only enthusiasts and the military made any serious use of the medium.

Today satellites are a part of everyday life, with TV delivered this way probably being the most visible aspect for the majority of people. The basic concept of satellite communications has been with us for well over half a century. It all began with Arthur C. Clarke's ground-breaking (space-breaking?) *Wireless World* article published in 1945 in which he showed that three satellites placed in geostationary orbit would have footprints covering about 90% of the Earth's surface. *Geostationary* means that the spacecraft's orbits and orbital speeds make them appear stationary with respect to the corresponding point on the Earth's surface immediately below the orbiting spacecraft.

Following World War II, European reconstruction and other, more fundamentally pressing needs demanded international attention. As a result, it was many years before technology reached the point where it became feasible to seriously consider satellite communications as a practical possibility.

In 1957, the Soviets launched the first artificial satellite, Sputnik, and the Space Race was well and truly on. On June 10, 1962, the first Telstar communications satellite was launched, paving the way for space-based international communications on a wider scale. This first-ever Telstar was placed into an unusual orbit. The spacecraft for this 1962 project had an elliptical orbit varying from 800-km to 4,800-km altitude, and only the relatively near-Earth segment, when the satellite was around 800–1,000 km up, was usable for communications. The Telstar satellites were eventually purchased by AT&T, and the name has been continued up to the present day.

This first Telstar operated on a bearer frequency of 4.17 GHz, and the transponder could either relay one (yes—just one) TV channel or alternatively 60 simultaneous phone conversations. This may all seem tame by modern standards, but the technology was truly revolutionary in 1962. Remember, there were no commercial integrated circuits available; it was when any low-powered computer would fill a large office suite, and even transistors were pushed to work at frequencies into VHF. The spacecraft for this program was designed and assembled by what was then Bell Labs, now operating as Nokia Bell Labs. However, the first Telstar proved that one could actually relay radio communications through an orbiting spacecraft and there followed various other experiments such as Project Relay and Early Bird.

Arthur C. Clarke's concept of the GEO and wide Earth coverage remained the linchpin that would lead to most of the progress in the 1970s and 1980s. Indeed, only a year after the first Telstar, the first GEOs were launched by NASA and Hughes and were known by the generic name Syncom. Nowadays, international trunk satellite communications, served by Intelsat and others, and also services like DTH and DBS (satellite TV) all use GEOs.

9.1.2 Low Earth Orbits and Geostationary Earth Orbits

Terrestrial communications networks such as 5G (see Chapter 8) are very limited in terms of reaching extensive rural regions such as many parts of Africa and India. Indeed:

- In 2023, 45% of Africans could not gain access to the internet, and, in Chad, that number is an appalling 98%.
- Due to sheer economics, groups of mainly young people typically crowd over just one connected device.
- Exceptions tend to be parts of Egypt and South Africa.
- It is important to bear in mind that the continent of Africa as a whole comprises a substantial number of sovereign states.

In contrast, the Indian subcontinent is of itself one sovereign state and, although a substantial proportion of the country is largely populated by relatively poor people, this nation has a proud and effective technological history, including space.

In these instances and many others, economics determine that terrestrial communications networks can never reach many rural communities, but SATCOM certainly can.

The main types of satellite orbits suitable for communications applications are shown schematically in Figure 9.1:

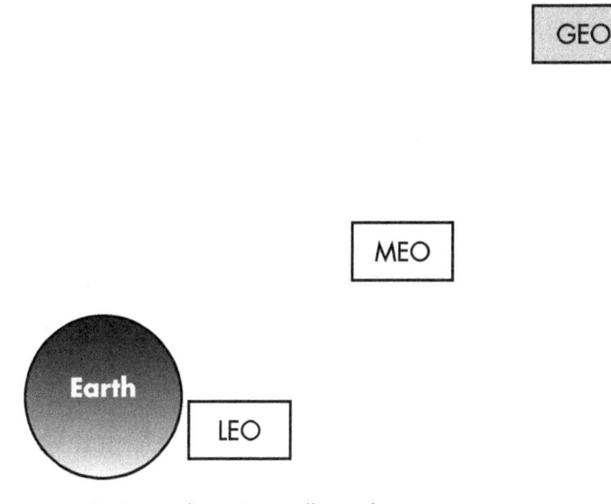

Figure 9.1 LEO, MEO, and GEO satellite orbits.

- *LEO:* 500 to 1,200-km range in practice;
- *MEO:* Medium Earth orbit, which is nominally 20,500 km in altitude and is usually an elliptical orbit;
- *GEO:* Necessarily placed specifically at the critical geostationary altitude of 35,786 km. GEOs orbit at this relatively high altitude, meaning a very substantial path length from Earth to satellite and a correspondingly substantial latency.

LEOs were historically largely avoided because of their relatively small area of Earth coverage. But by 2020, advances in communication capabilities had transformed the landscape so that LEO-based SATCOM is now highly effective and indeed competitive.

Current and prospective LEO constellations such as OneWeb or SpaceX all require in excess of 600 satellites in order to function satisfactorily.

In the 2020s, LEO constellation populations tend to be the approximate range: $600 \leq N \leq 50{,}000$.

Further details are provided below where the OneWeb and SpaceX Starlink systems are described.

There are also global positioning satellites (GPS, GNSS, GLONASS), circling at 20,200 km, considerably lower than GEOs because they must retain some relative motion in order to provide the navigational information from a multiple-satellite configuration.

Transmission delays with all the LEO systems are much shorter than for the GEOs, being mostly in the tens of milliseconds range. In contrast, GEO users suffer a round-trip transmission time delay of 540 ms, and this is very noticeable when using a GEO-based speech channel. Echo cancellers are regularly installed to eliminate echo in these types of links. GEO-based systems involve negligible Doppler shift problems because of their geostationary character. Although LEOs, in particular, have relatively small transmission time delays, the problem of Doppler shift is much greater than with GEOs because of the substantial differences between satellite motion and terrestrial receivers. It has only been practical to correct for this Doppler shift with the advent of advanced electronic systems. Link power budgets are greatly eased with LEO systems because of the much closer proximity to Earth, allowing all solid state technology to be implemented increasingly even at the higher-frequency bands, including mmWave.

9.1.3 Some Late Twentieth-Century Mobile Satellite System Projects

The concept of and designs for LEO-based satellite systems are certainly not new; indeed, in the late twentieth century, several projects attracted billion-dollar investments [1]. Because these projects provided substantial learning bases toward today's mature systems, we summarize two here: namely SkyBridge and Teledesic.

9.1.3.1 SkyBridge

It was planned for SkyBridge to have used the Ku-band, and it would have employed frequency reuse adaptively in order to offer real broadband services. Following these decisions, the utilization of the RF spectrum was also to be optimized as well as protecting GEO and terrestrial communications systems that were using the same frequencies.

SkyBridge (then an Alcatel Company) decided on the Ku-band because of the availability of proven and less expensive technology at these somewhat lower frequencies. Other features included:

- The avoidance of any intersatellite links (no ISLs);
- All switching was to be based in the ground-based control installations;
- The standard ATM protocol would be used (see Chapters 1 and 2).

The absence of ISLs combined with all the switching taking place on the ground was expected to increase the overall reliability and robustness of the system at the same time as keeping costs down below $4.2 billion. However, a drawback is that this approach limits connectivity and requires that all communications pass through hub Earth stations. SkyBridge was being designed to offer up to 20 Mbps to commercial customers and up to 2 Mbps to residential customers, thus representing a high-speed internet on-ramp capability. Any multiple of this capacity would have been made available to business users on demand and, overall, a total global market of over 20 million users was confidently expected. By 1998, a large industrial team was in place under the leadership of Alcatel to design and develop SkyBridge. More than 400 engineers were working on the program, and these large-scale engineering activities were expected to enable Alcatel to finalize the design characteristics of their system.

However, the project fell well short of the anticipated demand and was placed on hold in January 2002. Suffice to say, Skybridge never recovered sufficiently to regain any traction.

9.1.3.2 Teledesic

In the 1990s, when the Teledesic system was first proposed, it comprised a historically massive 840-satellite constellation, all with intelligent ISLs. By the mid-1990s, the constellation concept has been reduced by almost threefold down to 288 operational satellites, divided into 12 planes, each with 24 spacecraft.

The plan required the satellites to orbit in planes having north-to-south and south-to-north orientations, obviously with the Earth rotating beneath this network. Teledesic is probably the best known of all the early broadband satellite projects and there are two reasons for this. The first is probably just by being "first in," as described above. The second and very important reason is doubtless inextricably linked to the fact that none other than Microsoft's Bill Gates joined with Craig McCaw to initiate this venture. The contract team originally comprised Boeing (USA) and France's Matra but was later expanded to include Motorola. Figure 9.2 is an artist's impression of a Teledesic spacecraft, dominated by the large solar arrays for electrical power. The phased array antennas are facing Earth and therefore hidden from view in this picture.

The full constellation of 288 satellites would have looked like that depicted in Figure 9.1 (bottom left LEO in the illustration).

Interactive signals to and from the satellites would have been relayed between the network, mobile, or stationary ground and sea nodes; aircraft; and other spacecraft via ISLs. It was also considered possible in the future, given suitable gateways and handover protocols, that communications between other competing space-based networks could have been made available. The average latency (the time delay as the signal travels through the uplink and then the downlink) amounted to 70 ms with Teledesic, and the organization was anticipating that around 750,000 small-to-medium-sized businesses in the United States alone will go for these broadband services. At the end of the twentieth century, only 3% of such businesses had access to any level of fiber connection and the expectation was that this situation would only slowly change over the early years of the third millennium. Fiber connections were discussed in Chapters 3 and 6.

Figure 9.2 An artist's impression of a Teledesic spacecraft. (*From:* [18]. Courtesy of Teledesic LLC.)

As originally planned, the Teledesic system was to use Ka-band with a bearer frequency of 30 GHz for the uplinks and 20 GHz for the downlinks. Originally it was anticipated that Teledesic would become operational by 2000, but this became less likely when the management announced a funding program covering a period of 4 years. By 2002 or 2003, this advanced "space-based internet II" facility was expected to be functioning commercially.

Somewhat like Skybridge, the Teledesic system failed to generate anywhere near sufficient commercial interest and the management team shut down operations in 2002.

9.2 CURRENT AND PROJECTED MOBILE SATELLITE SYSTEMS

In the first edition of this title (written in 1999) the following statement was made regarding broadband MSS: "It is extremely unlikely that all or even most of the broadband satellite systems described in Section 9.2 will ever get built." As the economics turned out during the early years of the twenty-first century, it looks clear that none of those late twentieth-century hopefuls actually became a reality. Those courageous early developments paved the way for the new post-2020 MSS era. With modern LEOs, "seamless" 24-hour digital communications are maintained using relatively large constellations but also with greatly improved and relatively low-cost SpaceX-defined systems. Extremely fast communication links for spacecraft-to-ground and intersatellite links (ISL) are also very important advances. There has been and still is a steady decrease of the rocket launch costs for spacecraft.

The net result for MSS communications' commercial forecasted take-up is steady increases as shown in Figure 9.3.

It is anticipated global markets will more than double by 2029 and continue to grow thereafter. By far, the main driving force is the implementation of 5G NTN (nonterrestrial networks) which is a strongly internet-based feature with high-growth potential. As you read this book, the twenty-first-century new convergence of terrestrial communications and NTNs is progressing rapidly.

The development and exploitation of nanosatellites, starting in the late 1990s, has proved to be revolutionary. More recently (as the twenty-first century progressed), the term CubeSats has become the de facto term for extremely small LEO-based satellites (for some details, see Chapter 1). A CubeSat typically comprises a largely aluminum

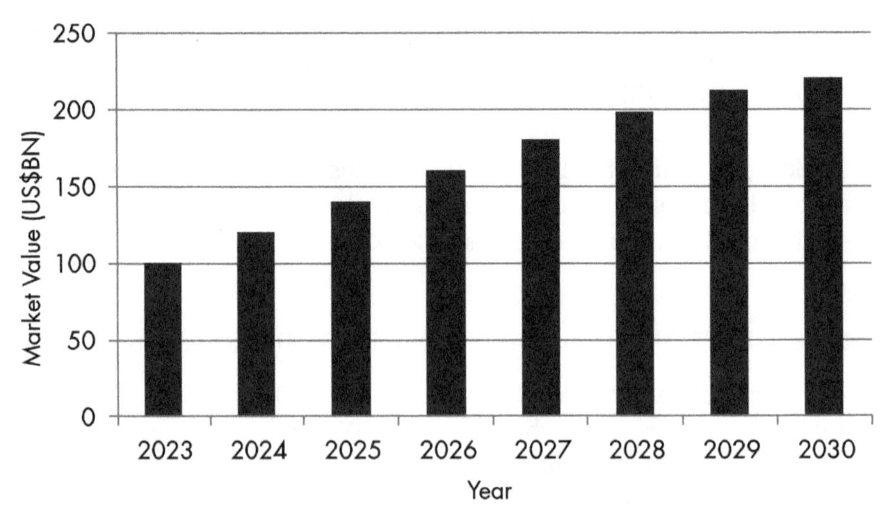

Figure 9.3 Total combined global markets for MSS (market values in billions of U.S. dollars). (The author's estimates.)

cubic structure having 10-cm sides into which all the electronics are packed. This means masses of digital chips including memory banks (basically a nano-computer), the entire RF system comprising the SDR, and the feed to the antennas. These days the actual antennas are almost always flat-panel phased arrays. For me this situation echoes my time working in the aerospace industry when my job was to progress microminiaturization while packing all the electronics into a military standard rack. For more information on CubeSats, type "CubeSats" into your search engine and check the Wikipedia entry which in my opinion provides a particularly good presentation with many interesting details. Then you can update with additional information as the industry develops further.

Regarding rocket launches of CubeSats Erik Kulu [2] has provided a very good column chart showing details dating from 1998 through to 2027. The year 1998 was 2 years before the first edition of this book was published and Kulu's chart indicates that, from that year through 2005, a total of just 22 such launches took place. For the present purpose, Kulu's basic data has been reused and some extra post-2023 data was added to generate Figure 9.4.

The purple curve tracks Kulu's data while the green curve includes additional data for 2023 onward. Factors leading to the difference in these suggested numbers are described by following Table 9.1.

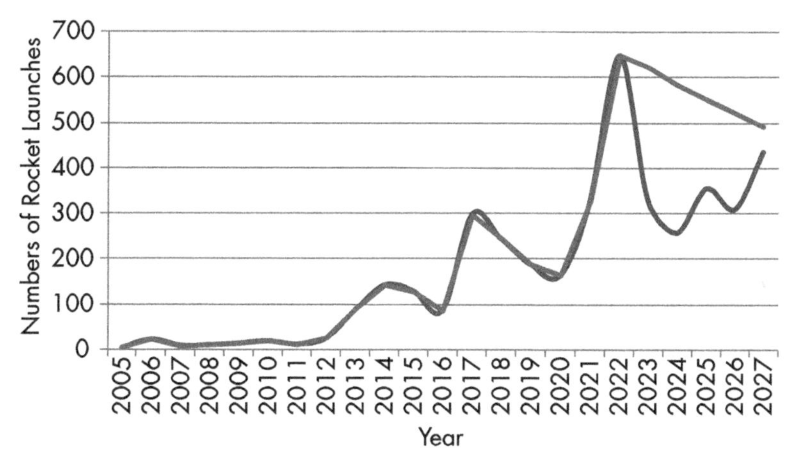

Figure 9.4 Numbers of rocket launches (carrying CubeSats) per annum 2005 through 2027.

Table 9.1
The Four Main LEO-Based Communications Satellite Constellations (MSS)

	SpaceX Starlink (Operating)	OneWeb (Operating)	Telesat Lightspeed	Amazon Project Kuiper
Future Constellation	Possibly as many as 42,000	2,000	1,600	3,236
Critical Number	1,440	648	298	578
RF Channel*	Ku, Ka, and E-bands	Ku-band	Ka-band	Ka-band
Altitude (km)†	560	1,200	1,000	590 to 630
Approximate Life Expectancy (Years)	5 to 7	5	10 to 12	7
Latency (ms)	<50	<50	<50	Not known
Major Aiming Markets	Cellular backhaul and consumer broadband	Cellular backhaul, consumer broadband, and government	Carrier-grade requirements, mobility, and government	Cellular backhaul and consumer broadband

Notes:
*Most currently envisioned MSS use either Ku-band or Ka-band.
†The quoted values for altitudes are nominal because each constellation is divided-up into several shells and planes.

At the time of this writing, there were four main contenders for commercial new MSS projects:

- SpaceX Starlink;
- OneWeb;
- Telesat Lightspeed;
- Amazon Project Kuiper.

Data points relating to each of these projects are summarized in Table 9.1 and described in detail next.

In all cases, these New Space MSS are clearly more viable than earlier (late 1990s/early 2000s) projected systems (see Section 9.1.3) due to the implementation of:

- Advanced modulation schemes such as 256QAM;
- Digital communications payloads;
- Multibeam antennas;
- Optimization through software-defined technology;

- Decreased launch costs, with a strong trend for further decreases.

The first two projects, OneWeb and SpaceX Starlink, were conceived as long ago as around 2000. Each project is now described in some detail. In each instance a table is provided summarizing the orbital parameters: where appropriate, phases or shells, orbits, orbital planes, and inclination angles.

9.2.1 OneWeb

This 644-satellite MSS project was completed late April 2023 and has been fully operational since that time [3, 4]. This is a notably special achievement considering that, in 2020, the OneWeb company was rescued from bankruptcy by the Indian company known as Bharti Global joining with the U.K. government. Since that year, OneWeb had raised a total $2.7 billion in order to complete its setting-up operations, principally to launch all 644 of the required satellites. In 2021, the company had its first 36 satellites waiting to be launched from Russia's Baikonur Cosmodrome. However, Russia's war against Ukraine posed an extremely serious issue and, unsurprisingly, Russia refused to even return any of OneWeb's spacecraft. The company required an alternative launch facility, which was eventually facilitated by the Satish Dhawan Space Centre in Sriharikota, India.

Details concerning OneWeb's constellation are given in Table 9.2.

9.2.2 SpaceX Starlink

Elon Musk's SpaceX (and other projects) is at least as well-known as OneWeb, although SpaceX was originally conceived and founded in

Table 9.2
Orbital Parameters for the OneWeb Constellation

Orbits (km Above Sea Level)	Phases	Numbers of Satellites	Orbital Inclinations (Degrees of Arc)
1,200 km (all)	Total 2	644	
	Phase 1	Two sets of 12 and 8 planes	87.9° and 55°
	Phase 2 (polar orbiting)	49 satellites per orbital plane	(Not known)

2002—a courageous decision bearing in mind the failures of the earlier MSS projects such as SkyBridge and Teledesic (see Section 9.1.3). Musk's observations regarding the decreasing cost of rocket launches coupled with the increasing digitization of the electronics and the vital prospects provided by CubeSats all led him to believe that there really were solid future business prospects in space. It was to be 10 years before SpaceX truly became a reality and the project actually went operational in 2022.

SpaceX's main technical specifications (satellites) are:

- Frequency bands: Ku, Ka, and E-band;
- Satellite antennas: phased arrays;
- Laser transponders on some spacecraft (this will likely increase in the future);
- Hall-effect thrusters on all spacecraft (the use of Hall-effect thrusters (electrical propulsion) means extremely rapidly maneuvering, which has already proved vital for spacecraft to avoid colliding with other celestial objects).

SpaceX Starlink's satellite constellations [3, 5] are segmented into five distinct "shells" (1 through 5), and details concerning these constellations (the data being from late 2022) are given in Table 9.3.

Adding together all the data in the third column of Table 9.3, the total number of Starlink satellites as of late 2022 was 3,808, but this keeps increasing as SpaceX manufactures (Redmond, Washington) and launches more satellites.

Table 9.3
Orbital Parameters for the SpaceX Starlink Constellations (Late 2022)

Orbits (Above Sea Level)	Shell	Numbers of Satellites	Orbital Inclinations (Degrees of Arc)
550 km	Shell 1	1,140	53°
540 km	Shell 2	1,440	53.2°
——	Shell 3	720	70°
560 km (Polar orbit)	Shell 4	336	97.6°
560 km (Polar orbit)	Shell 5	172	97.6°

Source: [6].

Clearly, from the presented information (including Table 9.3), this is already a huge project. Yet, according to Elon Musk, the founder of this project, it is set to grow even more massive.

By trawling the literature available on the internet, I obtained enough data to create the chart shown as Figure 9.5.

In spring 2023, there was a brief pause in the production of Starlink satellites, but thereafter the rate of manufacture resumed, albeit at a somewhat slower rate than late 2022. The projection is for a resumption in higher-rate production starting again in 2024.

Elon Musk already has the Federal Communications Commission (FCC) approval for starting up the second generation of SpaceX Starlink satellites (Gen2), which should allow him to increase the total number of spacecraft to an almost unprecedented 29,988. These satellites (plus several more) would be placed in new orbital shells as follows:

- First almost 10,000 orbiting at 525, 530, and 535 km;
- A further almost 20,000 orbiting at the considerably lower altitudes of 340 to 360 km;

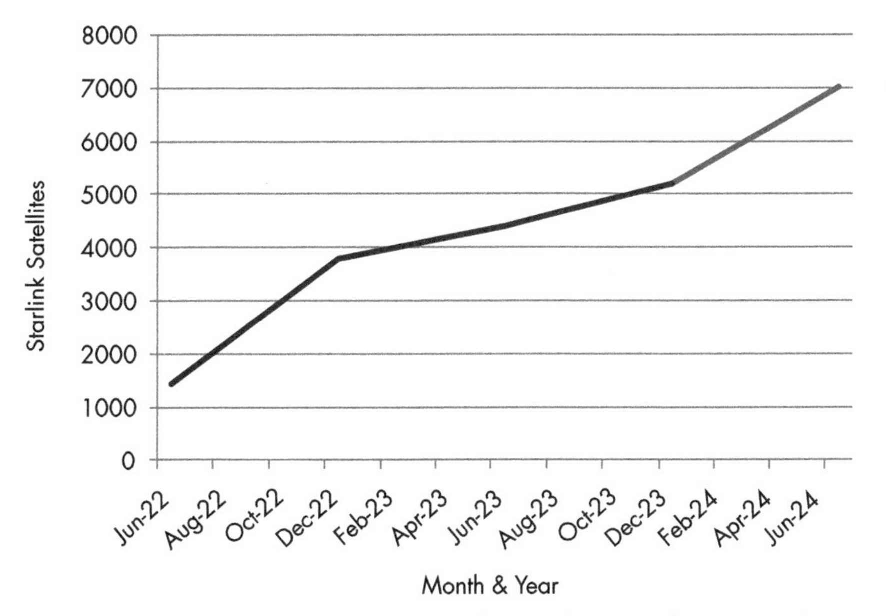

Figure 9.5 Numbers of Starlink satellites launched at specific times mid-2022 through mid-2024 (the blue curve refers to actual launches and the red curve refers to projected data).

- Another almost 500 at the considerably higher altitudes of 604 to 614 km.

By mid-2023, Starlink was serving over 1.5 million subscribers and counting.

This second generation could occur by around 2027 or 2028. Meanwhile, here are two notes of caution:

- Because of the increasing capacity of the individual rockets, the increases in the frequencies of rocket launches do not necessarily translate into proportionate increased numbers of satellites being launched.
- Incumbent satellite companies, including Eutelsat, Hughes Network Systems, SES, Telesat, and Viasat, have not just been sitting down and watching these new and highly active "kids on the block" (especially SpaceX) expand rapidly and take all. Starting in 2023, these incumbents were vigorously fighting back [7].

There are two further MSS contenders that must be outlined here: Amazon Project Kuiper and Telesat Lightspeed.

9.2.3 Amazon Project Kuiper

Almost needless to say, following Elon Musk and his SpaceX Starlink, another giant of the twenty-first century's internet-strong business environment, Jeff Bezos, has also ventured his company Amazon's LEO-based project, namely Amazon's Project Kuiper [8]. Amazon is very well known internationally for its warehouse-based internet shopping business.

Based on an overall concept somewhat similar to SpaceX Starlink, Bezos' Project Kuiper is aiming to launch 3,236 satellites into three orbital shells.

Details from late 2022 concerning Project Kuiper's constellations are given in Table 9.4.

The plan is for the system to operate at the Ka-band, which is right in the middle of either OneWeb or Starlink's bands.

The Kuiper Plan is for the project to comprise five phases of launches and deployment to complete the constellation and Amazon

Table 9.4
Orbital Parameters for the Project Kuiper Constellations in Late 2022: The First 3 Shells

Orbits (Above Sea Level)	Shell	Numbers of Orbital Planes and Contained Satellites
590 km	Shell 1	28 each with 28 satellites
610 km	Shell 2	36 each with 36 satellites
630 km	Shell 3	34 each with 34 satellites

Source: [3, 8]. No further data (including orbital inclinations) was available at the time of this writing.

aims to include coverage of the most densely populated regions of the Earth.

9.2.4 Telesat Lightspeed

Telesat is Canada-headquartered and was also cited in Chapter 7. Interestingly, the company is owned by the U.S. firm Loral Space & Communications. In fall 2023, Telesat's MSS project known as Lightspeed [9] was still in its concept stage and the main constellation parameters are shown in Table 9.5.

9.2.5 Future Prospects

9.2.5.1 The Overall Scenario

The four MSS projects identified and discussed above are OneWeb, Starlink, Project Kuiper, and Telesat Lightspeed. Of these projects, only OneWeb and Starlink are commercially operational and are therefore way out in front of any other contenders. The big question

Table 9.5
Orbital Parameters for the Telesat Lightspeed Constellations

Orbits (Above Sea Level)	Phases	Numbers of Satellites
1,015 km (polar) and 1,325 km (inclined)	Phase 1 (6 polar planes and 20 inclined planes)	13 satellites per orbital plane
Future evolution	Phase 2	39 satellites per orbital plane

No further data (including orbital inclinations) was available at the time of this writing. Like Project Kuiper, the plan is for the system to operate at the Ka-band, which is a higher-frequency band than either OneWeb or Starlink.

therefore is: Can Project Kuiper and/or Telesat Lightspeed possibly catch up with OneWeb and Starlink? Taking each of these in turn, Amazon expects its Project Kuiper to have launched its first prototype satellites around the end of 2023. You are reading this book in 2024 or later, so it will be very interesting to see if the companies have achieved this and indeed have accelerated spacecraft production and launch starting sometime in 2024.

There are perhaps three factors that could enable Project Kuiper and/or Telesat Lightspeed to compete effectively with the likes of Starlink:

- Greater capability in terms of satellite throughput;
- Better value for money for subscribers (e.g., greater bandwidth/price ratio than competitors);
- Substantially increased adoption of optical links.

Some or even all these and other objectives may be attainable given Amazon's multibillion-dollar financial investment capability.

Meanwhile, Telesat Lightspeed reckons that it will be 2026 before it launches its first satellites. By then, it is almost certain OneWeb and SpaceX Starlink, possibly also Project Kuiper, will surely have mature networks serving around 10 million or more customers well in place and expanding globally.

Could Telesat possibly gain enough traction to catch up with or even overtake the likes of OneWeb and Starlink? If Telesat Lightspeed actually gets off the ground, it could possibly be a winner in terms of its projected 10 to 11-year lifetime, which is about double most of the competition (see Table 9.1).

Regarding optical links, we know that at least some Starlink satellites already have laser transponders on board (see Section 9.2.2).

The expectation is that, within the next few years, the issue of scintillation will be solved (see Chapter 4) meaning that optical systems will become high-quality, practical, and acceptable. Maybe within a decade or so all networks will include at least some optical links? Perhaps projects such as Amazon Project Kuiper and/or Telesat Lightspeed will include such links, actually enabling them to catch up with or even overtake OneWeb and SpaceX Starlink?

9.3 TERRESTRIAL USER TERMINALS

Most end-users of MSS are ground-based customers desiring to take advantage of often being on the move and requiring broadband internet (5G, 6G) connection. During the era after 2023, the preferred antenna technology comprised flat-panel arrays [10, pp. 5–6]. Typical figures of merit for flat-panel receiver arrays designed for the consumer and enterprise broadband markets are [3]:

- G/T (receiver gain divided by its noise temperature) approximately 14–15 dB/K;
- EIRP (effective isotropic radiated power) 47 dBW.

SpaceX (http://www.starlink.com) is almost certainly the principal manufacturer of consumer-oriented, mobile, ground-based terminals, but there are several independent providers (offering interestingly distinct technologies) and a selection of these companies is now covered in this context.

9.3.1 ALCAN Systems GmbH

This company is based in Darmstadt, Germany, where it specializes in liquid-crystal (LC)-based smart antennas [11]. At the time of this writing, an LC-based flat-panel phased array was under development and production will be assembly-line-based.

9.3.2 All.Space

All.Space [12] operates in Reading, Berkshire, United Kingdom, where it manufactures its radome-based 5G smart MSS transceiver products. Effective August 1, 2022, the company previously known as Isotropic Systems rebranded as All.Space [13].

9.3.3 Hanwa Phasor

Hanwa Phasor [14] operates in Cambridge and London, United Kingdom. In June 2020, the Hanwa Group acquired the assets of Phasor Solutions (a London-based company) and, consequently, the new operation was named Hanwa Phasor.

The company claims their MSS ground-based terminal is a full active electronically scanning array (AESA) design. The product is

100% focused on LEO MSS systems and, at the time of this writing, procurement was through a company named Plexus.

9.3.4 Kymeta

Kymeta [15] employs over 500 people, most of whom work at the company's production factory in Seattle, Washington. This company operates a partnership with OneWeb and it caters for both a GEO and a LEO connection capability. The technical specifications of Kymeta's products include:

- Extended Ku-band coverage (10.7–12.75-GHz receiver and 13.75–14.50-GHz transmitter);
- Turnkey solutions for many applications with a built-in automatic control unit (ACU), tracking receiver plus options for transceiver and modem configurations;
- Designed for high agility when operating with a mobile user;
- Provides full-duplex operation from a single aperture;
- The patented antenna feed design enables the inclusion of a low-profile antenna with a high aperture efficiency;
- Fast beam pointing (the system reorientates within milliseconds), supporting LEO requirements.

9.3.5 SatixFy

Headquartered in Rehovot, Israel, SatixFy [16] employs a total of about 150 people. SatixFy also has either OneWeb or Starlink operations in Bulgaria, the United Kingdom, and the United States.

SatixFy manufactures ground-based (AESA) MSS terminals known as Amber and airborne units called Onyx Aero. They also supply satellite payloads to ESA and OneWeb.

9.3.6 ThinKom

ThinKom [17] has all its main operations in Hawthorne, California, United States, where it employs about 100 people. The company has developed an antenna system mainly for airborne MSS applications and this involves two antennas (one Ku-band and the other Ka-band) that cooperate so the user can select either band. ThinKom also implements variable inclination continuous transverse stub (VICTS) technology. Basically, the technology (Figure 9.6) comprises a set of lightweight discs rotating around a single axis to steer the antenna.

9.3.7 Flat-Panel Antennas or Dish Reflectors?

Most of the manufacturers cited above (Sections 9.3.1 through 9.3.6) implement flat-panel antennas. There exists a fundamental issue that can (and will) degrade performance, especially at low angles of elevation, which will be required with ground-based systems. The issue is caused by reflections from usually nearby ground surfaces and objects such as seating, bushes, trees, and people.

Flat-panel antennas are relatively open to such disturbances that can easily disrupt the signal path. A much better choice would almost always be the traditional dish reflector with the aperture at the feed point. Our TV receiver antennas are a good example of the principle. In effect, the aperture is then largely shielded from ground disturbances. In Chapter 8, it was pointed out that terrestrial wireless

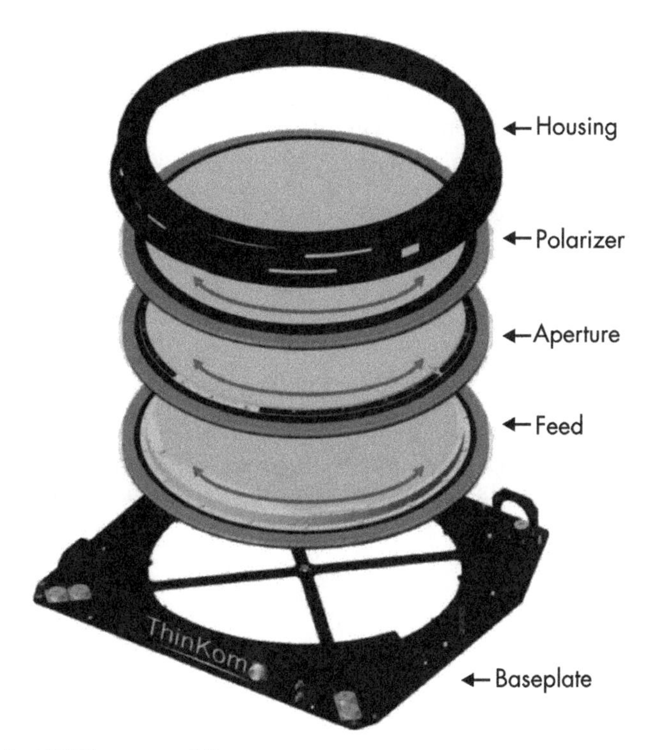

Figure 9.6 VICTS antenna [6].

antenna systems almost always use dish reflector technology. Almost all suppliers of radios for these applications cite the use of dish reflectors, for the important reasons of relatively high gain and negligibly low reflection issues.

References

[1] Liu, S., et al., "LEO Satellite Constellations for 5G and Beyond: How Will It Reshape Vertical Domains?" *Proc. IEEE*, April 2021.

[2] https://www.nanosats.eu/#figures.

[3] Correia, R., et al., "User Terminal Segments for Low-Earth Orbit Satellite Constellations," *IEEE Microwave Magazine*, October 2022, pp. 47–58.

[4] WorldVu Satellites Ltd., "International Bureau," Federal Communications Commission, Washington, D.C., Rep. No. SAT-MPL-2021011200007, 2022, https:licensing.fcc.gov/egi-bin/wx.exe/prod/ib/forms/reports/swr031b.hts?q_set=V_SITE_ANTENNA_FREQ.File_numberC/File+Number.

[5] https://www.spacex.com.

[6] Varrall, G., *5G and Satellite RF and Optical Integration*, Norwood, MA: Artech House, 2003, pp. 74–75, 176–183, and 196.

[7] https://interactive.satellitetoday.com/July-2023/traditional-operators-start-the-fightback.

[8] https://www.tomsguide.com/news/what-is-project-kuiper.

[9] https://www.telesat.com/leo-satellites.

[10] Edwards, T. C., *Technologies for RF Systems*, Norwood, MA: Artech House, 2018.

[11] www.alcansystems.com.

[12] https://www.all.space/products-services.

[13] https://spacenews.com/isotropic-rebrands-as-all-space.

[14] www.hanwha-phasor.com.

[15] www.kymetacorp.com.

[16] www.satixfy.com.

[17] www.thinkom.com.

[18] Edwards, T., *Gigahertz and Terahertz Technologies for Broadband Communications,* 1st ed., Norwood, MA: Artech House, 2000.

10

HIGH-SPEED DIGITAL EPILOGUE

10.1 INTRODUCTION TO THE FUTURE

It is January 16, 2040, and Betty Manton is at her home suffering the severe pain associated with gallstones. Twenty years ago, her mother Lois was in the same condition, but Lois had to wait 18 months before she could receive treatment in a hospital 60 miles from her home. Today Betty only has to await the arrival of 6G-connected and AI-augmented equipment at her home so the operation to remove her gall bladder can proceed on site. The requisite first-class standard of hygiene is automatically taken care of and the keyhole surgery is performed by the precision robot. Betty is continuously looked after extremely well throughout the operation as well as postoperatively.

This kind of scenario could actually be in place throughout much of the developed world well before the middle of the twenty-first century. Such vital medical procedures will be among many areas of human interventions and activities that require expert human support, although this is often greatly delayed at the time of this writing [1].

In pursuit of these kinds of goals, most of the technological advances described in this book will be required, plus:

- 6G;
- Supercomputers;

- Quantum technology;
- Powerful technological convergence.

These advances are described and discussed in this final chapter of the book. AI will also become increasingly pervasive (see Chapter 1).

10.2 THE 6G ERA

10.2.1 The Reality of 6G

Surely, 6G is a somewhat awkward and misleading term. In earlier chapters, we saw the sequence of mobile (cellular) communications is defined as: 2G, 3G, 4G, and 5G. It should be clear that not only does 5G sequentially follow after 4G, 5G has been and still is truly revolutionary because it embodies techniques that can only be realized by adopting digital approaches (e.g., software-defined networks, network slicing, and many other technological advantages).

In spite of being named 6G, this is actually a type of hugely important and yet challenging outgrowth of the already existing and ever-evolving 5G. The future was never 4G + 5G; it is 5G + 6G. Here is how Greg Jue of Keysight Technologies described 6G [2] (reprinted with permission from *Microwave Journal*):

> 6G aims to be the first generation of wireless technology to improve the quality of human life by bridging the physical, digital and mechanical worlds. Accomplishing this will mean adding artificial intelligence (AI) to networks to make them more efficient as well as building high fidelity digital twins. It will also require expanding spectrum use and building upon network architectures like nonterrestrial networks and highly virtualized disaggregated networks that began in 5G. For 6G to meet these goals, the spectrum allotted for wireless communications needs to be used more efficiently and new spectrum needs to be studied. Without expanding into new spectrum bands, it will be impossible to meet the high data throughput needs of applications like immersive telepresence, virtual reality, and extended reality.

Following several years of contemplation by around 2022, it became clear that new, extended, RF technologies would be required for 6G to cope with the immense bandwidth challenges of this emerging era. In an ideal world, it would be desirable for center frequencies around and even exceeding 1 THz (1,000 GHz) to be employed, but subterahertz frequencies showed early promise (e.g., frequencies in the 158 to 330-GHz range) [2, 3] at least for extremely broadband backhaul. Major global companies, notably Keysight Technologies (United States) and Rohde & Schwarz (Germany), are now highly active in this field.

10.2.2 Some 6G Technological Developments

Keysight Technologies, for example, have investigated the feasibility and performance of a quasi-optic test setup comprising an over-the-air (OTA) short-range subterahertz link [2]. This setup is shown in Figure 10.1.

The concept of using lenses for quasi-optic transmission is also outlined in Figure 10.1. The radiating element of the transmitting source comprises a diagonal feed horn antenna. The beam passes through a lens that causes it to converge at first to form what is known as the beam waist. Following this, the beam diverges ahead of entering the detection process. A more rigorous description of quasi-optic systems can be found in [3].

This Keysight Technologies subterahertz test bed was used to investigate an OTA measurement at 285 GHz with 30-GHz bandwidth for a point-to-point transmission carrying a signal approaching 100 Gbps, transmitting over a distance of 8m (26.5 ft). High-gain, high-directivity antennas such as phased array antennas are necessary to overcome free-space path loss. However, this was impossible because a 220 to 330-GHz phased array antenna was not available at the time of this experiment. Instead, quasi-optic techniques were investigated

Figure 10.1 Quasi-optic transmission, assuming identical feed horns for the source and receiver. (Reprinted with permission from *Microwave Journal* [2].)

using commercial off-the-shelf (COTS) lenses with performance specified to 220 GHz. For this experiment, the test bed was split between a transmit section and a receive section. A schematic diagram of the setup is shown in Figure 10.2.

Using 16-QAM modulation of the signal almost 100-Gbps transmission rate was achieved associated with a measured satisfactory EVM, but the range was limited to only 8m.

Full details regarding the RF components, subsystems, and instrumentation are provided in the article by Greg Jue [2].

Meanwhile, Rohde & Schwarz have been researching the propagation of subterahertz signals in both outdoor and street environments and have reported their results in detail for 158-GHz and 300-GHz signals. The results contributed to the ITU-R Working Party 5D report, *Technical Feasibility of IMT in Bands Above 100 GHz*. The report was consulted at the ITU World Radio Conference 2023 (WRC23) where additional frequency bands beyond 100 GHz were discussed and considered for allocation at WRC27. The earliest 6G standards are likely to be published around 2028.

Quasi-optic is described in [3] and several articles relating mainly to subterahertz technology (up to and through 100 Gbps) are provided in [4]. Tables of data, charts, and pictures of MMICs are provided in most of these articles. MMIC technology was outlined in Chapter 1.

However, during 2023, interest grew towards utilizing the midband (approximately 7–24 GHz) for 6G primary operating frequencies. This band is unofficially called FR3 and the reason for the renewed interest is the realization of how difficult it is to use even subterahertz frequencies. This applies particularly to handheld applications and

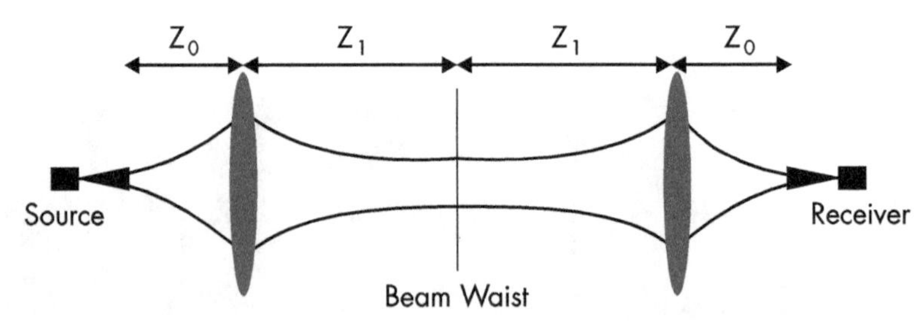

Figure 10.2 OTA quasi-optic transmission at 285 GHz with 30-GHz bandwidth, receive side. (Reprinted with permission from *Microwave Journal* [2].)

also more generally where large antennas cannot be utilized. In Chapter 8, there was a further reference to this issue.

It is important to appreciate most if not all current (early 2020s) work is at the research level. In a few years, developments will steadily become practical, increasingly orientated towards real commercially feasible systems.

10.2.3 The Importance of Metamaterials

Reconfigurable intelligent surfaces (RIS) implementing metamaterials are becoming essential features of the 6G high-bandwidth landscape [5]. What exactly are metamaterials? Perhaps the most significant aspect of these is they are not conventional materials. They are not elements such as copper, silver, and magnesium and they are not any type of ceramic or plastic either. Instead, they are combinations of metal, ceramic, and/or plastic engineered to form artificial structures that exhibit specific electromagnetic properties that are unavailable with conventional materials.

Metamaterials are designed to interact with electromagnetic radiation (RF or light) in a desired fashion. Information regarding the application of metamaterials to microwave guiding media is provided in [6, pp. 595–606].

Metamaterials generally comprise an array of structures within which the smallest element is smaller than the lowest wavelength within the applied electromagnetic radiation range. The meta-atoms interact with the electric and magnetic components of the radiation in a way that natural atoms do not. These metamaterials can be used to control electromagnetic radiation covering almost any frequency range, including negative refraction in the instance of optical radiation. They are also fundamental to reconfigurable intelligent surfaces [5].

10.2.4 Reconfigurable Intelligent Surfaces

Although this section focuses on 6G, the issues leading to considerations of reconfigurable intelligent surfaces also applies to 5G. This is all to do with the considerable difficulties associated with transmitting radio and light over substantial terrestrial distances (i.e., ranges). This challenge became clear in earlier chapters, Chapter 8 in particular. Traditionally, active repeaters are implemented enabling signals to ultimately cover up through tens of even hundreds of kilometers, but these repeaters come with several challenges that are inherent to the basic design. Such challenges include the limited efficiency of the RF power amplifiers necessarily used in the transmitter part of the microwave or millimeter-wave transmitters, and this limited efficiency inevitably results in the emission of heat into the transceiver structure. As the operating frequency increases, so amplifier efficiency decreases, which means proportionally even more heat escapes and eventually the temperature of the repeater can cause malfunction. At the subterahertz frequencies projected to be required for 6G, one notable question is: In practice, what technology may emerge to replace the repeaters?

It is worth observing again that for Keysight Technologies' demonstration setup, described above, the range achieved was only 8m, whereas, in practice, upwards of many kilometers will be required.

The implementation of RIS is one particularly promising approach that leads to signal regeneration without any need for repeaters. Researchers at Tsinghua University, Beijing, China, are notably active in this area of technology, which is described in [5]. The example shown in [5] is a square RIS comprising 2,304 unit cells (i.e., 48 × 48 cells within a square matrix). To illustrate the basic simplicity of this RIS concept, a very basic example of this approach is provided by the schematic grid of 12 × 12 cells shown in Figure 10.3.

A photograph of the actual 2,304-cell RIS is shown in [5, p. 43]. Operation is similar to that associated with an active phased array antenna. Each of the unit cells is controlled by a digital chip usually comprising a micro-controlled field-programmable gate array (FPGA). This also stores the coding sequences required to dynamically tune the entire RIS. The microcontroller sends the appropriate instructions, usually binary coding, to each unit cell, setting its state either to on or off. Cell switches are either varactor diodes or FETs, both of which are described in [7]. Most of the DC power is consumed by the microcontroller and the overall power consumption is generally around 12-W maximum, which is very much less than almost any conceivable active repeater. All of these associated electronics are attached to the rear of the RIS.

RIS is probably the only viable technology capable of coping with signal amplification at 6G subterahertz signals on a commercially viable basis. Further information is provided in [5].

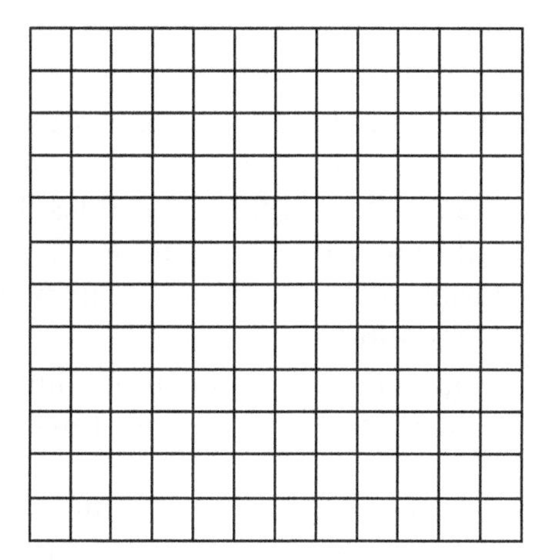

Figure 10.3 The basic (square matrix) layout of a simple 12 × 12 RIS. Each small square represents a unit cell.

10.3 SUPERCOMPUTERS

It may possibly surprise some, but the fact is that not all regular computers can solve any problem or numerical issue, not even top-of-the-range models. Subjects ranging as widely as cancer research and weather forecasting require the vastly greater computing capability available with either supercomputers or the radically different quantum computers. In this section, supercomputers are described, while quantum computers are covered in Section 10.4.

The rationale for including some technological and operational aspects of both supercomputers and quantum computers is simply that these types of systems lead to increasingly high-speed (highly broadband) signals, again demanding ever more bandwidth for processing and transmission. In an overall sense, supercomputers comprise input/output subsystems, interconnects, memory blocks, and processor cores. However, unlike our familiar desktop or laptop devices, supercomputers use several processing units (CPUs), and these are grouped into what are known as compute nodes, comprising a processor or a group of processors (called symmetric multiprocessing (SMP)) and a memory block.

A typical supercomputer contains some tens of thousands of nodes (as many as 50,000) and each node has interconnect capabilities so they can collaborate to solve a particular problem. These nodes also communicate with input/output subsystems such as data storage and networking.

A major concept required for any appreciation of supercomputers concerns “flops.” This does not refer to anything going wildly wrong; nothing is actually flopping. Neither does this term infer that some form of very informal footwear is under discussion (i.e., flip-flops). In computer terminology, this is an acronym meaning the number of floating-point operations per second.

For efficient programmability, operations are performed floating-point, and decimal points are included toward the end of the processing.

Even the computer on which I am typing does not have a specification that includes its flops; then, like the very best PC in the world, it is really very small by supercomputer standards.

A quarter-way into the twenty-first century, supercomputers are known for their flop capabilities that reach well into the petaflops realm, with top-ranking examples operating at exaflops. Peta (P) is standard mathematics for 10^{15}, while exa (E) represents 10^{18}. This means that a supercomputer functioning at 1 petaflop processes each operation a thousand trillion (1 quadrillion) times every second (i.e., each operation occupies 1 femtosecond of time). Such a supercomputer will possess vastly more processing power than the fastest laptop available.

As I write this book, there is just one truly exaflop-level supercomputer functioning (the Oak Ridge Frontier, described shortly), but, as you read this book, there may well be several additional exaflop machines.

The electrical power consumption of a typical supercomputer is in the order of some tens of kilowatts and this represents two specific downsides:

- Operating costs (including cooling systems);
- Acceptability in an increasingly environmentally critical world (i.e., sustainability).

So please do not plan soon for the delivery of your home-based supercomputer just yet, not even if you live right next door to Bill Gates or Elon Musk.

In 2023, the benchmark specifications of some then-leading supercomputers were as follows:

- World #1 was the Frontier system at the Oak Ridge National Laboratory in Tennessee, the United States. This system operates at 1.194 exaflops, is based on the latest Cray EX235a architecture, and is equipped with AMD EPYC 64C 2-GHz processors. The number of cores approaches 9 million, and the electrical efficiency is just over 52 gigaflops per watt (a gigaflop is one-millionth of a petaflop). An interconnect subsystem known as Sling-11 provides data transfer.
- World #2 is the Fugaku system at the RIKEN Center for Computational Science (R-CCS) in Kobe, Japan. This system has almost 8 million cores and operates at 442 petaflops.
- The globally third-ranking system is LUMI, which is a Cray EX installed at the EuroHPC Center in Kajaani, Finland. LUMI operates at 309.1 petaflops.

There is also the Leonardo supercomputer system installed at another EuroHPC site in CINECA, Italy, plus other systems located internationally that variously operate at some hundreds of petaflop levels. On a geographic basis, the leading countries are currently the United States, Japan, Finland, and Italy, but it would not be surprising to see countries such as Korea joining the list in the near future.

The United Kingdom's AI Supercomputer is probably the most powerful supercomputer under development. With up to 21-exaflop capability, this machine uses over 5,000 Green Hopper processors manufactured by Nvidia.

In 1976, the big news was the very fast Cray-1 installed at Los Alamos National Laboratory. Cray-1 operated at the blistering speed of about 160 Mflops (yes, mega-flops), and all of us celebrated this momentous achievement by top-class dining at that time. Nowadays we are well into the petaflop and exaflop era. Surely we can do even better, much better, harnessing a totally contrasting technology and architecture.

10.4 QUANTUM COMPUTERS

10.4.1 A Radical Shift in Thinking About Fundamentals

In terms of rapid advancement (adopting radically new technology), quantum computers will take some beating. In 2023, for example, China's Jinzhang quantum computer was operating at 180 million times the speed of the world's fastest supercomputer (i.e., the Frontier system at the Oak Ridge National Laboratory in Tennessee, which operates at 1.194 exaflops). Multiplying these two numbers shows the Jinzhang machine operates at a staggering 214.92 Yotta flops (1 Yotta flop = 10^{24} flops). Like almost all technology, computer operating speeds will continue increasing over the years ahead. So, what is so very different about quantum?

10.4.2 Quantum Physics Versus Classical Physics

Quantum physics comprises the fundamentals of quantum computer technology, and this area of physics is of an entirely different class compared with the physics underlying all electronics (classical physics). Classical physics underlies the great majority of the technologies described in this book (very much including semiconductors) until now. Quantum physics determines the behavior of photons and states of matter that cannot be included within classical physics.

Semiconductor electronics (sometimes termed solid-state electronics) is very much within the scope of classical physics, and it underlies the behavior of transistors, integrated circuits, and, indeed, the TWTs—all described in Chapter 1.

Quantum physics does not recognize electron behavior and instead is founded on the much more fundamental behavior of matter that is governed by discrete changes in energy, known as quanta. Quantum theory was developed in the early twentieth century when scientists, including Heisenberg and Planck, published their findings on the subject. There is only space for a brief introduction here, but we should start with the fact that Planck found the nice simple result for quantum energy (E_q):

$$E_q = hf \quad \text{(Joules)} \tag{10.1}$$

In this equation, Planck's constant $h = 6.626 \cdot 10^{-34}$ J.s and f is the frequency in hertz.

Obviously Planck's constant is a very small number, but it is interesting and informative to examine what happens when the frequency increases into the terahertz ranges (important with 6G; see Section 10.2). For example, at a frequency of 1 THz, applying (10.1) yields: $6.626 \cdot 10^{-34} \cdot 1 \cdot 10^{12} = 6.626 \cdot 10^{-22}$, which is a low yet appreciable amount of energy.

The importance of the quantum energy hf is particularly noticeable when compared with the thermal energy kT, which is controlled by Boltzmann's constant (k) and the temperature T as follows:

$$E_t = kT \text{ (Joules)} \tag{10.2}$$

The dimensionless ratio (10.1)/(10.2) is:

$$hf/kT \tag{10.3}$$

in which Boltzmann's constant $k = 1.38 \cdot 10^{-23}$ J.K^{-1}.

Again at 1 THz and at an ambient absolute temperature of 290K (T), (10.3) yields: $(6.626 \cdot 10^{-22}/1.38 \cdot 10^{-23} \cdot 290) = 0.1656$, which is a much more manageable quantity indicating the importance of the combined quantum and thermal energies.

Because (10.3) includes the operating temperature (T) in its denominator, it follows that decreasing the temperature increases the ratio and this, in turn, makes the quantum term more prominent. At the same time, thermal noise is reduced, which is important because quantum systems are particularly sensitive to noise.

Equation (10.1) forms a basis for quantum physics, whereas, in contrast, (10.2) is important in classical physics (particularly semiconductor theory). In his book [8], Geoff Varrall well described the closely related and important security technique known as quantum key distribution (QKD) [8, pp. 178–180].

10.4.3 Qubits and Quantum Computers

Every digital computer fundamentally requires the generation and manipulation of binary digits or bits, as these are generally known. In all electronic digital computers, the bits are generated and manipulated electrically with a predetermined voltage representing a digital 1 and a different voltage (often close to 0V) representing a digital 0.

For a quantum computer, we again need a device that can be recognized in either one of two states representing a digital 1 or a digital 0. This is provided by a quantum bit, which is almost always known as a qubit. Just as an electronically processed bit is the basis of a conventional electronic computer (classical physics), so a qubit is the basis of a quantum computer. Then things become more complex. A qubit is a basic two-state quantum-mechanical system that functions according to the laws of quantum mechanics. In particular, there are the following quantum characteristics of an electron and a photon that can be used to represent a qubit: electron spin and photon polarization.

Electron spin-up could be chosen to represent digital 1 and then spin-down would represent 0.

Horizontal photon polarization could be chosen to represent digital 1; then vertical polarization would represent 0. Like electrons in electronic computers of any kind, qubits can be combined to form logic gates. The big difference being the resulting quantum computers are hundreds of millions of times faster than their wholly electronic predecessors. A basic block diagram showing the main functional elements of a quantum computer is shown in Figure 10.4.

In Figure 10.4, the output digital words are just representative (i.e., not necessarily the real outputs for any particular processing sequence).

State maintenance duration is a serious issue with qubits because, in most instances, this duration is typically only just over 100

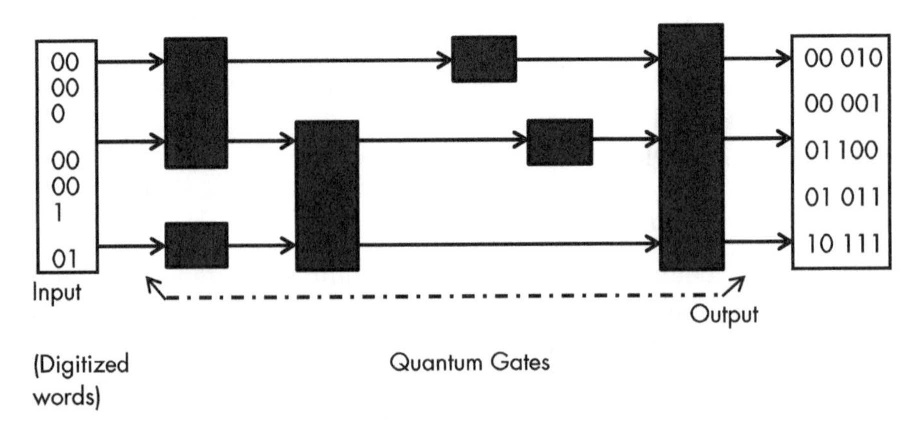

Figure 10.4 Basic elements of a quantum computer.

microseconds. However, in 2009, Michael Devoret (Yale University) discovered that fluxonium qubits maintain their quantum states for at least 1.48 ms, which represents a very substantial improvement [9]. In September 2023, this technology was taken up commercially as announced by D-Wave Quantum, Inc.

The basic functions of electronic digital gates are well known and are summarized in [7, Section 5.4.1]. For an electronic computer, these gates are usually manufactured in digital CMOS chip technology and are almost always realized using NAND gates (as an example, all the technology in USB memory sticks is generally flash-NAND).

For a quantum computer, all the quantum gates could still be NAND, but these have to be triggered on or off using, for example, the different spin states of electrons or the polarization of photons.

10.4.4 Some Basic Technologies Concerning Quantum Computers

There are many underlying technologies for quantum computers that have emerged over several years and are the subjects of current and future research. Because of its importance, we now focus on cryogenic technology in this role. The quantum advantage of operating at relatively low temperatures was described in Section 10.4.2.

Some quantum computers require operation in a superconductive environment and, in such configurations, a structurally simple device known as a Josephson Junction (JJ) is required.

The Josephson Effect only operates under superconductive conditions (i.e., cryogenic, less than 1K). Under this effect, a continuous current flows even when there is no applied voltage and such a phenomenon is completely foreign in an electronic context. This is an example of a physical phenomenon that can only occur under quantum conditions.

The Josephson Junction is also known as a 3D Transmon Qubit, which is manufactured using an electron beam source to deposit aluminum conductor strips and magnetron sputtering of the niobium; this entire structure forms a junction that occupies about a 1-micron square typically deposited on a quartz substrate. The term Transmon Qubit suggests the necessary concept of a qubit that can be used for signal processing.

In order to ensure that the signal amplitude is sufficient for quantum processing, it must be increased in level, and this requires a device known as a parametric amplifier or paramp, in this case, a Josephson-based paramp (JPA). Other important requirements include:

- The qubit must be immersed in a resonator.
- This structure must be isolated from the rest of the system.

To achieve this quantum amplifier, a passive microwave device known as a circulator must be interposed, as shown in Figure 10.5.

Symbolically, this entire quantum amplifier (G) is shown in Figure 10.6. For example, when $G = 20$ dB and the entire amplifier is held at a temperature of 20 mK (i.e., 0.02K), the overall noise figure is 0.002156 dB, which is vastly lower than almost any conventional electronics and is vital for achieving and maintaining quantum conditions. The noise figure of the JPA is very small because there are no active components and the amplifier is held at an extremely low temperature, as defined above. This general kind of technology lies at the heart of the strongly growing field of quantum computers.

10.5 HYPERCONVERGENCE

As early as the 1970s, the overall prospect of technological convergence was being seriously considered. This meant the convergence of computers and communications networks, which is very much matter of a fact these days (it also underlies much of the material considered in this book). It is well worthwhile recollecting that true convergence

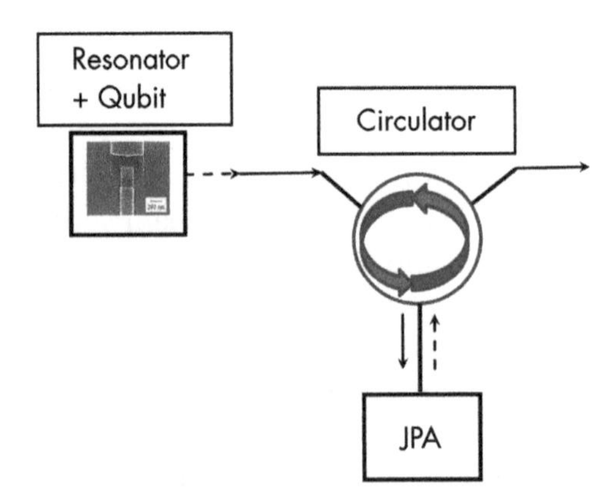

Figure 10.5 JPA with surrounding circuitry.

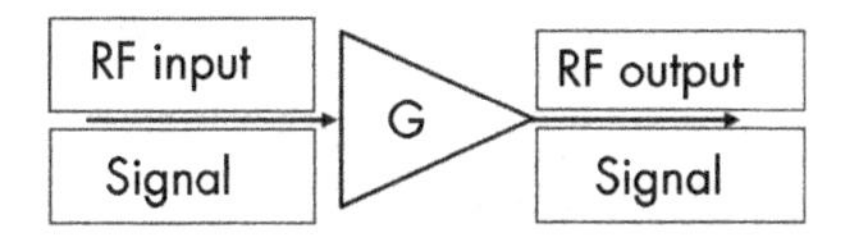

Figure 10.6 Schematic of the quantum amplifier with power gain *G*.

required communications with relatively wide bandwidths and the emergence of as least moderately powerful desktop computers, none of which really happened until the 1980s (Chapter 1). The proliferation of affordable and yet powerful digital devices running in conjunction with wide bandwidth fiber-optic cables, free-space optics, terrestrial wireless, HAPS, and satellite communications continues to drive increasing convergence. However, the Shannon limit regarding maximum bit rates is likely to remain in place.

In this final chapter, we have looked at supercomputers, quantum technology, and 6G. These and many other factors will be driving what we might call twenty-first-century hyperconvergence. It is imperative to recognize this dynamic and to ensure, as 5G advances and 6G begins its journey, all practicable effort will (via well-thought-through standards) remain in place as our world takes advantage of these exciting technological developments. The implementation of AI will likely be a substantial benefit, although great care will be required, particularly where higher levels of AI are being considered (see Chapter 1).

Referring to the beginning of this chapter hopefully we can expect many more "Betty Mantons" plus a plethora of further benefits arising from advanced applications of 6G, AI, optical systems, quantum computing, and NTN converged with terrestrial communications.

References

[1] whitepapers@go.electronics.com, "A Game-Changer For Wireless Clinical Devices: The Impact of Wi-Fi6.6E on Connected Medical Devices in Hospital Settings," announced September 15, 2023.

[2] Jue, G., "A Quasioptic OTA Transmission at 285 GHz with 30 GHz Bandwidth," *Microwave Journal*, March 2023, pp. 20–28.

[3] Goldsmith, P. F., *Quasioptical Systems: Gaussian Beam Quasioptical Propagation and Applications*, New York: Wiley-IEEE Press, 1998.

[4] Carlowitz, C., and M. Dietz, "Integrated Front-End Approaches for Wireless 100 Gb/s and Beyond," *IEEE Microwave Magazine*, August 2023, pp 16–34.

[5] Poulakis, M., "6G's Metamaterials Solution," *IEEE Spectrum*, November 2022, pp. 41–45.

[6] Edwards, T., and M. Steer, *Foundations for Microstrip Circuit Design*, 4th ed., New York: John Wiley & Sons, 2016.

[7] Edwards, T., *Technologies for RF Systems*, Norwood, MA: Artech House, 2018.

[8] Varrall, G., *5G and Satellite RF and Optical Integration*, Norwood, MA: Artech House, 2023.

[9] www.eenewseurope.com/en/high-dimensional-quantum-states, September 11, 2023.

ABOUT THE AUTHOR

Since 1989, **Terry Edwards** has headed-up a consultancy firm known as Engalco-Research, focusing on display systems as well as signal transmission technologies and the global industry. He holds an MPhil postgraduate degree in microwave research, has led seminars on fiber optics, and has written several articles, research papers, and books including:

- *Foundations for Microstrip Circuit Design* [John Wiley & Sons, First Edition, 1981; Second Edition, 1991; Third and Fourth Editions (together with Professor Michael Steer, NCSU), 2000 and 2016].
- *Microwave Electronics* (Edward Arnold, 1984).
- *Fiber Optic Systems—Network Applications* (John Wiley & Sons, 1989).
- *Gigahertz and Terahertz Technologies for Broadband Communications* (Artech House, First Edition, 2000).
- *Technologies for RF Systems* (Artech House, First Edition, 2018).

During the 1970s Terry taught postgraduate level microwave electronics at La Trobe University in Australia and since 1985 has led many consultancy projects, including for NXP Semiconductors (Netherlands). The NXP project comprised a major private-client twenty-first century research contract. He has also researched and prepared many industry/market reports focused on his specialist areas of technology. Under sponsorship with a North American corporation, he has reported on markets for microwave-based systems in Asia, Africa, and Latin America. Terry has travelled considerably to conferences and trade exhibitions in cities including Atlanta, Denver, Geneva, Munich, Paris, San Diego, San Jose, and Santa Rosa. Later in the 1990s he acted as an expert witness at a Philadelphia court hearing involving a fiber-optics product technology dispute.

In October 2021, by teleconference, Terry led a CEI-Europe organized 5-day seminar on noise in electronic systems.

Over the years Terry has been actively involved in country hiking, cycling, jogging, and some tennis.

INDEX

Artech House Mobile Communications Library

William Webb, Series Editor

3G CDMA2000 Wireless System Engineering, Samuel C. Yang

3G Multimedia Network Services, Accounting, and User Profiles, Freddy Ghys, Marcel Mampaey, Michel Smouts, and Arto Vaaraniemi

5G and Satellite RF and Optical Integration, Geoff Varrall

5G-Enabled Industrial IoT Networks, Amitava Ghosh, Rapeepat Ratasuk, Simone Redana, and Peter Rost

5G New Radio: Beyond Mobile Broadband, Amitav Mukherjee

5G Spectrum and Standards, Geoff Varrall

802.11 WLANs and IP Networking: Security, QoS, and Mobility, Anand R. Prasad and Neeli R. Prasad

Achieving Interoperability in Critical IT and Communications Systems, Robert I. Desourdis, Peter J. Rosamilia, Christopher P. Jacobson, James E. Sinclair, and James R. McClure

Advances in 3G Enhanced Technologies for Wireless Communications, Jiangzhou Wang and Tung-Sang Ng, editors

Advances in Mobile Information Systems, John Walker, editor

Advances in Mobile Radio Access Networks, Y. Jay Guo

Artificial Intelligence in Wireless Communications, Thomas W. Rondeau and Charles W. Bostian

Broadband Wireless Access and Local Network: Mobile WiMax and WiFi, Byeong Gi Lee and Sunghyun Choi

CDMA for Wireless Personal Communications, Ramjee Prasad

CDMA RF System Engineering, Samuel C. Yang

CDMA Systems Capacity Engineering, Kiseon Kim and Insoo Koo

Cell Planning for Wireless Communications, Manuel F. Cátedra and Jesús Pérez-Arriaga

Cellular Communications: Worldwide Market Development, Garry A. Garrard

Cellular Mobile Systems Engineering, Saleh Faruque

Cognitive Radio Interoperability through Waveform Reconfiguration, Leszek Lechowicz and Mieczyslaw M. Kokar

Cognitive Radio Techniques: Spectrum Sensing, Interference Mitigation, and Localization, Kandeepan Sithamparanathan and Andrea Giorgetti

The Complete Wireless Communications Professional: A Guide for Engineers and Managers, William Webb

Designing RF Combining Systems for Shared Radio Sites, Ian Graham

EDGE for Mobile Internet, Emmanuel Seurre, Patrick Savelli, and Pierre-Jean Pietri

Emerging Public Safety Wireless Communication Systems, Robert I. Desourdis, Jr., et al.

From LTE to LTE-Advanced Pro and 5G, Moe Rahnema and Marcin Dryjanski

The Future of Wireless Communications, William Webb

Geospatial Computing in Mobile Devices, Ruizhi Chen and Robert Guinness

Gigahertz and Terahertz Technologies for Broadband Communications, Second Edition, Terry Edwards

GPRS for Mobile Internet, Emmanuel Seurre, Patrick Savelli, and Pierre-Jean Pietri

GSM and Personal Communications Handbook, Siegmund M. Redl, Matthias K. Weber, and Malcolm W. Oliphant

GSM Networks: Protocols, Terminology, and Implementation, Gunnar Heine

GSM System Engineering, Asha Mehrotra

Handbook of Land-Mobile Radio System Coverage, Garry C. Hess

Handbook of Mobile Radio Networks, Sami Tabbane

Handbook of Next-Generation Emergency Services, Barbara Kemp and Bart Lovett

High-Speed Wireless ATM and LANs, Benny Bing

Implementing Full Duplexing for 5G, David B. Cruickshank

In-Band Full-Duplex Wireless Systems Handbook, Kenneth E. Kolodziej, editor

Inside Bluetooth Low Energy, Second Edition, Naresh Gupta

Interference Analysis and Reduction for Wireless Systems, Peter Stavroulakis

Interference and Resource Management in Heterogeneous Wireless Networks, Jiandong Li, Min Sheng, Xijun Wang, and Hongguang Sun

Internet Technologies for Fixed and Mobile Networks, Toni Janevski

Introduction to 3G Mobile Communications, Second Edition, Juha Korhonen

Introduction to 4G Mobile Communications, Juha Korhonen

Introduction to Communication Systems Simulation, Maurice Schiff

Introduction to Digital Professional Mobile Radio, Hans-Peter A. Ketterling

An Introduction to GSM, Siegmund M. Redl, Matthias K. Weber, and Malcolm W. Oliphant

Introduction to Mobile Communications Engineering, José M. Hernando and F. Pérez-Fontán

Introduction to OFDM Receiver Design and Simulation, Y. J. Liu

An Introduction to Optical Wireless Mobile Communications, Harald Haas, Mohamed Sufyan Islim, Cheng Chen, and Hanaa Abumarshoud

Introduction to Radio Propagation for Fixed and Mobile Communications, John Doble

Introduction to Wireless Local Loop, Broadband and Narrowband, Systems, Second Edition, William Webb

IS-136 TDMA Technology, Economics, and Services, Lawrence Harte, Adrian Smith, and Charles A. Jacobs

Location Management and Routing in Mobile Wireless Networks, Amitava Mukherjee, Somprakash Bandyopadhyay, and Debashis Saha

LTE Air Interface Protocols, Mohammad T. Kawser

Metro Ethernet Services for LTE Backhaul, Roman Krzanowski

Mobile Data Communications Systems, Peter Wong and David Britland

Mobile IP Technology for M-Business, Mark Norris

Mobile Satellite Communications, Shingo Ohmori, Hiromitsu Wakana, and Seiichiro Kawase

Mobile Telecommunications Standards: GSM, UMTS, TETRA, and ERMES, Rudi Bekkers

Mobile-to-Mobile Wireless Channels, Alenka Zajić

Mobile Telecommunications: Standards, Regulation, and Applications, Rudi Bekkers and Jan Smits

Multiantenna Digital Radio Transmission, Massimiliano Martone

Multiantenna Wireless Communications Systems, Sergio Barbarossa

Multi-Gigabit Microwave and Millimeter-Wave Wireless Communications, Jonathan Wells

Multiuser Detection in CDMA Mobile Terminals, Piero Castoldi

OFDMA for Broadband Wireless Access, Slawomir Pietrzyk

Practical Wireless Data Modem Design, Jonathon Y. C. Cheah

The Practitioner's Guide to Cellular IoT, Cameron Kelly Coursey

Prime Codes with Applications to CDMA Optical and Wireless Networks, Guu-Chang Yang and Wing C. Kwong

Quantitative Analysis of Cognitive Radio and Network Performance, Preston Marshall

QoS in Integrated 3G Networks, Robert Lloyd-Evans

Radio Resource Management for Wireless Networks, Jens Zander and Seong-Lyun Kim

Radiowave Propagation and Antennas for Personal Communications, Third Edition, Kazimierz Siwiak and Yasaman Bahreini

RDS: The Radio Data System, Dietmar Kopitz and Bev Marks

Resource Allocation in Hierarchical Cellular Systems, Lauro Ortigoza-Guerrero and A. Hamid Aghvami

RF and Baseband Techniques for Software-Defined Radio, Peter B. Kenington

RF and Microwave Circuit Design for Wireless Communications, Lawrence E. Larson, editor

Sample Rate Conversion in Software Configurable Radios, Tim Hentschel

Signal Failure: The Rise and Fall of the Telecoms Industry, John Polden

Signal Processing Applications in CDMA Communications, Hui Liu

Signal Processing for RF Circuit Impairment Mitigation, Xinping Huang, Zhiwen Zhu, and Henry Leung

Smart Antenna Engineering, Ahmed El Zooghby

Software-Defined Radio for Engineers, Travis F. Collins, Robin Getz, Di Pu, and Alexander M. Wyglinski

Software Defined Radio for 3G, Paul Burns

Software Defined Radio: Theory and Practice, John M. Reyland

Spectrum Wars: The Rise of 5G and Beyond, Jennifer A. Manner

Spread Spectrum CDMA Systems for Wireless Communications, Savo G. Glisic and Branka Vucetic

Technical Foundations of the IoT, Boris Adryan, Dominik Obermaier, and Paul Fremantle

Technologies and Systems for Access and Transport Networks, Jan A. Audestad

Third-Generation and Wideband HF Radio Communications, Eric E. Johnson, Eric Koski, William N. Furman, Mark Jorgenson, and John Nieto

Third Generation Wireless Systems, Volume 1: Post-Shannon Signal Architectures, George M. Calhoun

Traffic Analysis and Design of Wireless IP Networks, Toni Janevski

Transmission Systems Design Handbook for Wireless Networks, Harvey Lehpamer

UMTS and Mobile Computing, Alexander Joseph Huber and Josef Franz Huber

Understanding Cellular Radio, William Webb

Understanding Digital PCS: The TDMA Standard, Cameron Kelly Coursey

Understanding WAP: Wireless Applications, Devices, and Services, Marcel van der Heijden and Marcus Taylor, editors

Universal Wireless Personal Communications, Ramjee Prasad

Virtualizing 5G and Beyond 5G Mobile Networks, Larry J. Horner, Kurt Tutschku, Andrea Fumagalli, and ShunmugaPriya Ramanathan

WCDMA: Towards IP Mobility and Mobile Internet, Tero Ojanperä and Ramjee Prasad, editors

Wi-Fi 6: Protocol and Network, Susinder R. Gulasekaran and Sundar G. Sankaran

Wireless Communications in Developing Countries: Cellular and Satellite Systems, Rachael E. Schwartz

Wireless Communications Evolution to 3G and Beyond, Saad Z. Asif

Wireless Intelligent Networking, Gerry Christensen, Paul G. Florack, and Robert Duncan

Wireless LAN Standards and Applications, Asunción Santamaría and Francisco J. López-Hernández, editors

Wireless Sensor and Ad Hoc Networks Under Diversified Network Scenarios, Subir Kumar Sarkar

Wireless Technician's Handbook, Second Edition, Andrew Miceli

For further information on these and other Artech House titles, including previously considered out-of-print books now available through our In-Print-Forever® (IPF®) program, contact:

Artech House
685 Canton Street
Norwood, MA 02062
Phone: 781-769-9750
Fax: 781-769-6334
e-mail: artech@artechhouse.com

Artech House
16 Sussex Street
London SW1V 4RW UK
Phone: +44 (0)20 7596-8750
Fax: +44 (0)20 7630-0166
e-mail: artech-uk@artechhouse.com

Find us on the World Wide Web at: www.artechhouse.com